James Hall Nasmyth, James Carpenter

The Moon

Considered as a Planet, a World and a Satellite

James Hall Nasmyth, James Carpenter

The Moon
Considered as a Planet, a World and a Satellite

ISBN/EAN: 9783337367237

Printed in Europe, USA, Canada, Australia, Japan

Cover: Foto ©berggeist007 / pixelio.de

More available books at **www.hansebooks.com**

THE MOON:

CONSIDERED AS

A PLANET, A WORLD, AND A SATELLITE.

By JAMES NASMYTH, C.E.

AND

JAMES CARPENTER, F.R.A.S.

LATE OF THE ROYAL OBSERVATORY, GREENWICH.

WITH TWENTY-FOUR ILLUSTRATIVE PLATES OF LUNAR OBJECTS, PHENOMENA, AND SCENERY; NUMEROUS WOODCUTS, &c.

SECOND EDITION.

LONDON:
JOHN MURRAY, ALBEMARLE STREET.
1874.

TO

HIS GRACE THE DUKE OF ARGYLL,

IN RECOGNITION OF HIS LONG CONTINUED INTEREST IN THE
SUBJECT OF WHICH IT TREATS,

This Volume

IS MOST RESPECTFULLY DEDICATED

BY

THE AUTHORS.

PREFACE.

THE reason for this book's appearance may be set forth in a few words. A long course of reflective scrutiny of the lunar surface with the aid of telescopes of considerable power, and a consequent familiarity with the wonderful details there presented, convinced us that there was yet something to be said about the moon, that existing works on Astronomy did not contain. Much valuable labour has been bestowed upon the topography of the moon, and this subject we do not pretend to advance. Enough has also been written for the benefit of those who desire an acquaintance with the intricate movements of the moon in space; and accordingly we pass this subject without notice. But very little has been written respecting the moon's physiography, or the causative phenomena of the features, broad and detailed, that the surface of our satellite presents for study. Our observations had led us to some conclusions, respecting the cause of volcanic energy and the mode of its action as manifested in the characteristic craters and other eruptive phenomena that abound upon the moon's surface. We have endeavoured to explain these phenomena by reference to a few natural laws, and to connect them with the general hypothesis of planet formation which is now widely accepted by cosmologists. The principal aim of our work is to lay these proffered explanations before the students and admirers of astronomy and science in general; and we trust that what we have deduced concerning the moon may be taken as referring to a certain extent to other planets.

Some reflections upon the moon considered as a world, in reference to questions of habitability, and to the peculiar conditions which would attend a sojourn on the lunar surface, have appeared to us not inappropriate. These, though instructive, are rather curious than important. More worthy of respectful consideration are the few remarks we have offered upon the moon as a satellite and a benefactor to the inhabitants of this Earth.

In reference to the Illustrations accompanying this work, more especially those which represent certain portions of the lunar surface as they are revealed by the aid of powerful telescopes, such as those which we employed in our scrutiny, it is proper that we should say a few words here on the means by which they have been produced.——

During upwards of thirty years of assiduous observation, every favourable opportunity has been seized to educate the eye not only in respect to comprehending the general character of the moon's surface, but also to examining minutely its marvellous details under every variety of phase, in the hope of rightly understanding their true nature as well as the causes which had produced them. This object was aided by making careful drawings of each portion or object when it was most favourably presented in the telescope. These drawings were again and again repeated, revised, and compared with the actual objects, the eye thus advancing in correctness and power of appreciating minute details, while the hand was acquiring, by assiduous practice, the art of rendering correct representations of the objects in view. In order to present these Illustrations with as near an approach as possible to the absolute integrity of the original objects, the idea occurred to us that by translating the drawings into models which, when placed in the sun's rays, would faithfully reproduce the lunar effects of light and shadow, and then photographing the models so treated, we should produce most faithful representations of the original. The result was in every way highly satisfactory, and has yielded pictures of the details of the lunar surface such as we feel every confidence in submitting to those of our readers who have made a special study of the subject. It is hoped that those also who have not had opportunity to become intimately acquainted with the details of the lunar surface, will be enabled to become so by aid of these Illustrations.

In conclusion, we think it desirable to add that the photographic illustrations above referred to are printed by well-established pigment processes which ensure their entire permanency.

CONTENTS.

CHAPTER I.
ON THE COSMICAL ORIGIN OF THE PLANETS OF THE SOLAR SYSTEM.

Origination of Material Things—Celestial Vapours—Nebulæ—Their vast Numbers—Sir W. Herschel's Observations and Classification—Buffon's Cosmogony—Laplace's Nebular Hypothesis—Doubts upon its Validity—Support from Spectrum Analysis 1

CHAPTER II.
THE GENERATION OF COSMICAL HEAT.

Conservation of Force—Indestructibility of Force—Its Convertibility into Heat—Dawn of the Doctrine—Mayer's Deductions—Joule's Experiments—Mechanical Equivalent of Heat—Gravitation the Source of Cosmical Heat—Calculations of Mayer and Helmholtz—The Moon as an Incandescent Sphere—Not necessarily Burning—Loss of Heat by Radiation—Cooling of External Crust—Commencement of Selenological History 11

CHAPTER III.
THE SUBSEQUENT COOLING OF THE IGNEOUS BODY.

Cooling commenced from Outer Surface—Contraction by Cooling—Expansion of Molten Matter upon Solidification—Water not exceptional—Similar Behaviour of Molten Iron—Floating of Solid on Molten Metal—Currents in a Pot of Molten Metal—Bursting of Iron Bottle by Congelation of Bismuth within—Evidence from Furnace Slag—From the Crater of Vesuvius—Effects of Contraction of Moon's Crust and Expansion of Interior—Production of Ridges and Wrinkles—Theory of Wrinkles—Examples from shrivelled Apple and Hand 19

CHAPTER IV.
THE FORM, MAGNITUDE, WEIGHT, AND DENSITY OF THE LUNAR GLOBE.

Form of Moon—Not perfectly Spherical—Bulged towards Earth—Diameter—Angular Measure—Linear Measure—Parallax of Moon—Distance—Area of Lunar Sphere—Solid Contents—Mass of Moon—Law of Gravitation—Mass determined by Tides

and other Means—Density—How obtained—Specific Gravity of Lunar Matter—
Force of Gravity at Surface—How determined—Weights of similar Bodies on
Earth and Moon—Effects of like Forces acting against Gravity on Earth and
Moon 31

CHAPTER V.

ON THE EXISTENCE OR NON-EXISTENCE OF A LUNAR ATMOSPHERE.

Subject of Controversy—Phenomena of Terrestrial Atmosphere—No Counterparts on
Moon—Negative Evidence from Solar Eclipses—No Twilight on Moon—Evidence
from Spectrum Analysis—From Occultations of Stars—Absence of Water or
Moisture—Cryophorus — No Reddening of Sun's Rays by Vapours on Moon—
No Air or Water to complicate Discussions of Lunar Volcanic Phenomena . . 39

CHAPTER VI.

THE GENERAL ASPECT OF THE LUNAR SURFACE.

Pre-Telescopic Ideas—Human Countenance—Other supposed Resemblances—Portrait
of Full Moon—Permanence of Features—Rotation of Moon—Solar Period and
Solar Day on Moon—Libration—Diurnal—In Latitude—In Longitude—Visible
and Invisible Hemispheres — Telescopic Scrutiny—Galileo's Views — Features
Visible with Low Power—Low Powers on small and large Telescopes—Salient
Features—Craters—Plains—Bright Streaks—Mountains—Higher Telescopic
Powers—Detail Scrutiny of Features therewith—Discussion of High Powers—
Education of Eye—Highest practicable Power—Size of smallest Visible Objects . 51

CHAPTER VII.

TOPOGRAPHY OF THE MOON.

Reasons for Mapping the Moon—Early Maps—Labours of Langreen—Hevelius—
Riccioli—Cassini—Schroeter—Modern Maps—Lohrman's—Beer and Maedler's—
Excellence of the last—Measurement of Mountain Heights—Need of a Picture
Map—Formation of our own—Skeleton Map—Table of conspicuous Objects—
Descriptions of special Objects—Copernicus—Gassendi—Eudoxus and Aristotle—
Triesnecker—Theophilus, Cyrillus, and Catharina—Thebit—Plato—Valley of
the Alps — Pico—Tycho—Wargentin—Aristarchus and Herodotus—Walter—
Archimedes and the Apennines 65

CHAPTER VIII.

ON LUNAR CRATERS.

Use of term Crater for Terrestrial and Lunar Formations—Truly Volcanic Nature of
Lunar Craters—Terrestrial and Lunar Volcanic Areas compared—Similarity—
Difference only in Magnitude—Central Cone—Found in great and small Lunar
Craters—Formative Process of Terrestrial Volcanoes—Example from Vesuvius—
Vast Size of Lunar Craters—Reasons assigned—Origin of Moon's Volcanic

CONTENTS. xiii

PAGE

Force — Aqueous Vapour Theory untenable — Expansion upon Solidification Theory—Formative Process of a Lunar Crater—Volcanic Vent—Commencement of Eruption—Erection of Rampart—Hollowing of Crater—Formation of Central Cone—Of Plateau—Various Heights of Plateaux—Coneless Craters—Filled-up Craters—Multiple Cones—Craters on Plateau—Double Ramparts—Landslip Terraces—Rutted Ramparts—Overlapping and Superposition of Craters—Source-Connection of such—Froth-like Aggregations of Craters—Majestic Dimensions of Larger Craters 89

CHAPTER IX.

ON THE GREAT RING-FORMATIONS NOT MANIFESTLY VOLCANIC.

Absence of Central Cones—Vast Diameters—Difficult of Explanation—Hooke's Idea—Suggested Cause of True Circularity—Scrope's Hypothesis of Terrestrial Tumescences—Rozet's Tourbillonic Theory—Dana's Ebullition Theory . . 117

CHAPTER X.

PEAKS AND MOUNTAIN RANGES.

Paucity of extensive Mountain Systems on Moon—Contrast with Earth—Lunar Mountains found in less disturbed Regions—Lunar Apennines, Caucasus, and Alps—Valley of Alps—"Crag and Tail" Contour—Isolated Peaks—How produced—Analogy from Freezing Fountain—Terrestrial Counterparts and their Explanation by Scrope—Blowing Cone on Teneriffe—Comparative Gentleness of Mountain-forming Action—Relation between Mountain Systems and Crater Systems—Wrinkle Ridges 124

CHAPTER XI.

CRACKS AND RADIATING STREAKS.

Description—Divergence from Focal Craters—Experimental Explanation of their Cause—Radial Cracking of Crust—Outflow of Matter therefrom—Analogy from "Starred" Ice—No Shadows cast by Streaks—Their probable Slight Elevation—Open Cracks—Great Numbers—Length—Depth—In-fallen Fragments—Shrinkage a Cause of Cracks—Lateness of their Production . . . 133

CHAPTER XII.

COLOUR AND BRIGHTNESS OF LUNAR DETAILS: CHRONOLOGY OF FORMATIONS, AND FINALITY OF EXISTING FEATURES.

Absence of Conspicuous Colour—Slight Tints of "Seas"—Cause—Probable Variety of Tints in small Patches—Diversity of Brightness of Details—Most Conspicuous at Full Moon—Classification of Shades—Exaggerated Contrasts in Photographs—Brightest Portions probably the latest formed—Chronology of Formations—Large Craters older than Small—Mountains older than Craters—Bright Streaks comparatively recent — Cracks most recent of all Features — Question of existing Change — Evidence from Observation — Paucity of such Evidence —

xiv   CONTENTS.

                                                                        PAGE
Supposed Case of *Linné*—Theoretical Discussion —Relative Cooling Tendencies
of Earth and Moon—Earth nearly assumed its final Condition—Moon probably
cooled Ages upon Ages ago—Possible slight Changes from Solar Heating—Disintegrating Action   143

CHAPTER XIII.

THE MOON AS A WORLD: DAY AND NIGHT UPON ITS SURFACE.

Existence of Habitants on other Planets—Interest of the Question—Conditions of
Life—Absence of these from Moon—No Air or Water and intense Heat and
Cold—Possible Existence of Protogerms of Life—A Day on the Moon imagined
—Instructiveness of the Realization—Length of Lunar Day—No Dawn or
Twilight—Sudden Appearance of Light—Slowness of Sun in Rising—No Atmos-
pheric Tints—Blackness of Sky and Visibility of Stars and fainter Luminosities
at Noon-day—Appearance of the Earth as a Stationary Moon—Its Phases—
Eclipse of Sun by Earth—Attendant Phenomena—Lunar Landscape—Height
essential to secure a Point of View—Sunrise on a Crater—Desolation of Scene
—No Vestige of Life—Colour of Volcanic Products—No Atmospheric Perspective
—Blackness of Shadows—Impressions on other Senses than Sight—Heat of
Sun untempered—Intense Cold in Shade—Dead Silence—No Medium to conduct
Sound—Lunar Afternoon and Sunset—Night—The Earth a Moon—Its Size,
Rotation, and Features—Shadow of Moon upon it—Lunar Night-Sky—Con-
stellations—Comets and Planets—No Visible Meteors—Bombardment by Dark
Meteoric Masses—Lunar Landscape by Night—Intensity of Cold   155

CHAPTER XIV.

THE MOON AS A SATELLITE: ITS RELATION TO THE EARTH AND MAN.

The Moon as a Luminary—Secondary Nature of Light-giving Function—Primary
Office as a Sanitary Agent—Cleansing Effects of the Tides—Tidal Rivers and
Transport thereby—The Moon a "Tug"—Available Power of Tides—Tide-
Mills—Transfer of Tidal Power Inland—The Moon as a Navigator's Guide—
Longitude found by the Moon—Moon's Motions—Discovered by Observations—
Grouped into Theories—Represented by Tables—The Nautical Almanac—The
Moon as a Long-Period Timekeeper—Reckoning by "Moons"—Eclipses the
Starting-Points of Chronologies—Furnish indisputable Dates—Solar Surround-
ings revealed by Eclipses when Moon screens the Sun—Solar Corona—Moon
as a Medal of Creation, a Half-formed World—Abuses of the Moon—Super-
stitions — Erroneous Ideas regarding Moonlight betrayed by Artists and
Authors—The Moon and the Weather—Errors and Facts—Atmospheric Tides—
Warmth from Moon—Paradoxical Effect in cooling the Earth   171

CHAPTER XV.

CONCLUDING SUMMARY   184

LIST OF PLATES.

PLATE		PAGE
GASSENDI	*Frontispiece*	
I.—SUMMIT OF VESUVIUS		26
II.—WRINKLED HAND AND APPLE		30
III.—FULL MOON PHOTOGRAPH		52
IV.—PICTURE-MAP OF THE MOON	} *To face each other.*	68
V.—SKELETON MAP		
VI.—TERRESTRIAL AND LUNAR VOLCANIC AREAS COMPARED		88
VII.—PROGRESSIVE SERIES OF CRATERS		92
VIII.—COPERNICUS		96
IX.—THE LUNAR APENNINES, &c., &c.		100
X.—ARISTOTLE AND EUDOXUS		104
XI.—TRIESNECKER		108
XII.—THEOPHILUS, CYRILLUS, AND CATHARINA		112
XIII.—ARZACHAEL, PTOLEMY, AND THE RAILWAY		116
XIV.—PLATO, THE VALLEY OF THE ALPS, PICO, &c.		120
XV.—MERCATOR AND CAMPANUS		124
XVI.—TYCHO AND ITS SURROUNDINGS		128
XVII.—WARGENTIN		132
XVIII.—ARISTARCHUS AND HERODOTUS		136

LIST OF PLATES.

PLATE	PAGE
XIX.—FULL MOON AND CRACKED GLASS GLOBE, ILLUSTRATING THE CAUSE OF THE BRIGHT RADIATING STREAKS	140
XX.—OVERLAPPING CRATERS	148
XXI.—LUNAR CRATER. IDEAL LANDSCAPE	156
XXII—SOLAR ECLIPSE AS IT WOULD BE SEEN FROM THE MOON	164
XXIII—GROUP OF MOUNTAINS. IDEAL LUNAR LANDSCAPE	170

THE MOON.

CHAPTER I.

ON THE COSMICAL ORIGIN OF THE PLANETS OF THE SOLAR SYSTEM.

In this Chapter we propose to treat briefly of the probable formation of the various members of the solar system from matter which previously existed in space in a condition different from that in which we at present find it—*i.e.*, in the form of planets and satellites.

It is almost impossible to conceive that our world with its satellite, and its fellow worlds with their satellites, and also the great centre of them all, have always, from the commencement of time, possessed their present form: all our experiences of the working of natural laws rebel against such a supposition. In every phenomenon of nature upon this earth—the great field from which we must glean our experiences and form our analogies—we see a constant succession of changes going on, a constant progression from one stage of development to another taking place, a perpetual mutation of form and nature of the same material substance occurring: we see the seed transformed into the plant, the flower into the fruit, and the ovum into the animal. In the inorganic world we witness the operation of the same principle; but, by reason of their slower rate of progression, the changes there are manifested to us rather by their resulting effects than by their visible course of operation. And when we consider, as we are obliged to do, that the same laws work in the greatest as well as the smallest processes of nature, we are compelled to believe in an antecedent state of existence of the matter that composes the host of heavenly bodies, and amongst them the earth and its attendant moon.

In the pursuit of this course of argument we are led to inquire whether there exists in the universe any matter from which planetary bodies could be formed, and how far their formation from such matter can be explained by the operation of known material laws.

Before the telescope revealed the hidden wonders of the skies, and brought its rich fruits into our garner of knowledge concerning the nature of the universe, the philosophic minds of some early astronomers, Kepler and Tycho Brahe to wit, entertained the idea that the sun and the stars— the suns of distant systems—were formed by the condensation of celestial vapours into spherical bodies ; Kepler basing his opinion on the phenomena of the sudden shining forth of new stars on the margin of the Milky Way. But it was when the telescope pierced into the depths of celestial space, and brought to light the host of those marvellous objects, the nebulæ, that the strongest evidence was afforded of the probable validity of these suppositions. The mention of " nebulous stars " made by the earlier astronomers refers only to clusters of telescopic stars which the naked eye perceives as small patches of nebulous light ; and it does not appear that even the nebula in Andromeda, although so plainly discernible as to be often now-a-days mistaken by the uninitiated for a comet, was known, until it was discovered by means of a telescope, in 1612, by Simon Marius, who described it as resembling a candle shining through semi-transparent horn, as in a lantern, and without any appearance of stars. Forty years after this date Huygens discovered the splendid nebula in the sword handle of Orion, and in 1665 another was detected by Hevelius. In 1667 Halley (afterwards Astronomer Royal) discovered a fourth ; a fifth was found by Kirsch in 1681, and a sixth by Halley again in 1714. Half a century after this the labours of Messier expanded the list of known nebulæ and clusters to 103, a catalogue of which appeared in the " Connaissance du Temps " (the French " Nautical Almanac ") for the years 1783-1784. But this branch of celestial discovery achieved its most brilliant results when the rare penetration, the indomitable perseverance, and the powerful instruments of the elder Herschel were brought to bear upon it. In the year 1779 this great astronomer began to search after nebulæ with a seven-inch reflector, which he subsequently superseded by the great one of forty feet focus and four feet aperture. In 1786 he published his first catalogue of 1000 nebulæ; three years later he astonished the learned world by a second catalogue containing 1000 more, and in 1802 a third came forth comprising other 500, making 2500 in all! This number has been so far increased by the labours of more recent astronomers that the last complete catalogue, that of Sir John Herschel, published a few years ago, contains the places of 5063 nebulæ and clusters.

At the earlier periods of Herschel's observations, that illustrious observer appears to have inclined to the belief that all nebulæ were but remote clusters of stars, so distant, so faint, and so thickly agglomerated as to affect the eye only by their combined luminosity, and at this period of the nebular history it was supposed that increased telescopic power would resolve them into their component stars. But the familiarity which Herschel gained with the phases of the multitudinous nebulæ that passed in review before his eyes, led him ultimately to adopt the opinion, advanced by previous philosophers, that they were composed of some vapoury or elementary matter out of which, by the process of condensation, the heavenly bodies were formed; and this led him to attempt a classification of the known nebulæ into a cosmical arrangement, in which, regarding a chaotic mass of vapoury matter as the primordial state of existence, he arranged them into a series of stages of progressive development, the individuals of one class being so nearly allied to those in the next that, to use his own expression, not so much difference existed between them "as there would be in an annual description of the human figure were it given from the birth of a child till he comes to be a man in his prime." (*Philosophical Transactions, Vol. CI.*, pp. 271 *et seq.*)

His category comprises upwards of thirty classes or stages of progression, the titles of a few of which we insert here to illustrate the completeness of his scheme.

Class 1. Of extensive diffused nebulosity. (A table of 52 patches of such nebulosity actually observed is given, some of which extend over an area of five or six square degrees, and one of which occupies nine square degrees.)

„ 6. Of milky nebulosity with condensation.

Class 15. Of nebulæ that are of an irregular figure.
  ,, 17. Of round nebulæ.
  ,, 20. Of nebulæ that are gradually brighter in the middle.
  ,, 25. Of nebulæ that have a nucleus.
  ,, 29. Of nebulæ that draw progressively towards a period of final condensation.
  ,, 30. Of planetary nebulæ.
  ,, 33. Of stellar nebulæ nearly approaching the appearance of stars.

In a walk through a forest we see trees in every stage of growth, from the tiny sapling to the giant of the woods, and no doubt can exist in our minds that the latter has sprung from the former. We cannot at a passing glance discern the process of development actually going on; to satisfy ourselves of this, we must record the appearance of some single tree from time to time through a long series of years. And what a walk through a forest is to an observer of the growth of a tree, a lifetime is to the observer of changes in such objects as the nebulæ. The transition from one state to another of the nebulous development is so slow that a lifetime is hardly sufficient to detect it. Nor can any precise evidence of change be obtained by the comparison of drawings or descriptions of nebulæ at various epochs, with whatever care or skill such drawings be made, for it will be admitted that no two draughtsmen will produce each a drawing of the most simple object from the same point of view, in which every detail in the one will coincide exactly with every detail in the other. There is abundant evidence of this in the existing representations of the great nebula in Orion; a comparison of the drawings that have been lately made of this object, with the most perfect instruments and by the most skilful of astronomical draughtsmen, reveals varieties of detail and even of general appearance such as could hardly be imagined to occur in similar delineations of one and the same subject; and any one who himself makes a perfectly unbiassed drawing at the telescope will find upon comparison of it with others that it will offer many points of difference. The fact is that the drawing of a man, like his penmanship, is a personal characteristic, peculiar to himself, and the drawings of two persons cannot be expected to coincide any more than their handwritings. The appearance of a nebula varies also to a great extent with the power of the telescope used to observe it and the conditions under which it is observed; the drawings of nebulæ made with the inferior telescopes of a century or two centuries ago, the only ones that, by comparison with those made in modern times, could give satisfactory evidence of changes of form or detail, are so rude and imperfect as to be useless for the purpose, and it is reasonable to suppose that those made in the present day will be similarly useless a century or two hence. Since then we can obtain no evidence of the changes we must assume these mysterious objects to be undergoing, *ipso facto*, by observation of *one nebula* at *various periods*, we must for the present accept the *primâ facie* evidence offered (as in the case of the trees in a forest) by the observation of *various nebulæ* at *one period*.

"The total dissimilitude," says Herschel at the close of the observations we have alluded to, "between the appearance of a diffusion of the nebulous matter and of a star, is so striking, that an idea of the conversion of the one into the other can hardly occur to any one who has not before him the result of the critical examination of the nebulous system which has been displayed in this [his] paper. The end I have had in view, by arranging my observations in the order in which they have been placed, has been to show that the above-mentioned extremes may be connected by such nearly allied intermediate steps, as will make it highly probable that every succeeding state of the nebulous matter is the result of the action of gravitation upon it while in a foregoing one, and by such steps the successive condensation of it has been brought up to the planetary condition. From this the transit to the stellar form, it has been shown, requires but a very small additional compression of the nebulous matter."

Where the researches of Herschel terminated those of Laplace commenced. Herschel showed how a mass of nebulous matter so diffused as to be scarcely discernible might be and probably was, by the mere action of gravitation, condensed into a mass of comparatively small dimensions when viewed in relation to the immensity of its primordial condition. Laplace demonstrated how the known laws of gravitation could and probably did from such a partially condensed mass of matter produce an entire planetary system with all its subordinate satellites.

The first physicist who ventured to account for the formation of the various bodies of our solar system was Buffon, the celebrated French naturalist. His theory, which is fully detailed in his renowned work on natural history, supposed that at some period of remote antiquity the sun existed without any attendant planets, and that a comet having dashed obliquely against it, ploughed up and drove off a portion of its body sufficient in bulk to form the various planets of our system. He suggests that the matter thus carried off "at first formed a torrent the grosser and less dense parts of which were driven the farthest, and the densest parts, having received only the like impulsion, were not so remotely removed, the force of the sun's attraction having retained them:" that "the earth and planets therefore at the time of their quitting the sun were burning and in a state of liquefaction;" that "by degrees they cooled, and in this state of fluidity they took their form." He goes on to say that the obliquity of the stroke of the comet might have been such as to separate from the bodies of the principal planets small portions of matter, which would preserve the same direction of motion as the principal planets, and thus would form their attendant satellites.

The hypothesis of Buffon, however, is not sufficient to explain all the phenomena of the planetary system; and it is imperfect, inasmuch as it begins by assuming the sun to be already existing, whereas any theory accounting for the primary formation of the solar system ought necessarily to include the origination of the most important body thereof, the sun itself. Nevertheless, it is but due to Buffon to mention his ideas, for the errors of one philosophy serve a most useful end by opening out fields of inquiry for subsequent and more fortunate speculators.

Laplace, dissatisfied with Buffon's theory, sought one more probable, and thus was led to the proposition of the celebrated *nebular hypothesis* which bears his name, and which, in spite of its disbelievers, has never been overthrown, but remains the only probable, and, with our present knowledge, the only possible explanation of the cosmical origin of the planets of our system. Although Laplace puts forth his conjectures, to use his own words, "with the deference which ought to inspire everything that is not a result of observation and calculation," yet the striking coincidence of all the planetary phenomena with the conditions of his system gives to those conjectures, again to use his modest language, "a probability strongly approaching certitude."

Laplace conceived the sun to have been at one period the nucleus of a vast nebula, the attenuated surrounding matter of which extended beyond what is now the orbit of the remotest planet of the system. He supposed that this mass of matter in process of condensation possessed a rotatory motion round its centre of gravity, and that the parts of it that were situated at the limits where centrifugal force exactly counterbalanced the attractive force of the nucleus were abandoned by the contracting mass, and thus were formed successively a number of rings of matter concentric with and circulating around the central nucleus. As it would be improbable that all the conditions necessary to preserve the stability of such rings of matter in their annular form could in all cases exist, they would break up into masses which would be endued with a motion of rotation, and would in consequence assume a spheroidal form. These masses, which hence constituted the various planets, in their turn condensing, after the manner of the parent mass, and abandoning their outlying matter, would become surrounded by similarly concentric rings, which would break up and form the satellites surrounding the various planetary masses; and, as a remarkable exception to the rule of the instability of the rings and their consequent breakage, Laplace cited the case of Saturn surrounded by his rings as the only instances of unbroken rings that the whole system offers us; unless indeed we include the zodiacal light, that cone of hazy luminosity that is frequently seen streaming from our luminary shortly before and after sunset, and which Laplace supposed to be formed of molecules of matter, too volatile to unite either with themselves or with the planets, and which must hence circulate about the sun in the form of a nebulous ring, and with such an appearance as the zodiacal actually presents.

This hypothesis, although it could not well be refuted, has been by many hesitatingly received, and for a reason which was at one time cogent. In the earlier stages of nebular research it was clearly seen, as we have previously remarked, that many of the so-called nebulæ, which appeared at first to consist of masses of vapoury matter, became, when scrutinised with telescopes of higher power, resolved into clusters containing countless numbers of stars, so small and so closely agglomerated, that their united lustre only impressed the more feeble eye as a faint nebulosity; and as it was found that each accession of telescopic power increased the numbers of nebulæ that were thus resolved, it was thought that every nebula would at some period succumb to the greater penetration of more powerful instruments; and if this were the case, and if no real nebulæ were hence found to exist, how, it was argued, could the nebular hypothesis be maintained? One of the most important nebulæ bearing upon this question was the great one in the sword handle of Orion, one of the grandest and most conspicuous in the whole heavens. On account of the brightness of some portions of this object, it seemed as though it ought to be readily resolvable, supposing all nebulæ to consist of stars, but all attempts to resolve it were in vain, even with the powerful telescopes of Sir John Herschel and the clear zenethal sky of the Cape of Good Hope. At length the question was thought to be settled, for upon the completion of Lord Rosse's giant reflector, and upon examination of the nebula with it, his lordship stated that there could be little, if any, doubt as to its resolvability, and then it was maintained, by the disbelievers in the nebular theory, that the last stronghold of that theory had been broken down.

But the truths of nature are for ever playing at hide and seek with those who follow them:—the dogmas of one era are the exploded errors of the next. Within the past few years a new science has arisen that furnishes us with fresh powers of penetration into the vast and secret laboratories of the universe; a new eye, so to speak, has been given us by which we may discern, by the mere light that emanates from a celestial body, something of the chemical elements of which it is composed. When Newton two hundred years ago toyed with the prism he bought at Stourbridge fair, and projected its pretty rainbow tints upon the wall, his great mind little suspected that that phantom riband of gorgeous colours would one day be called upon to give evidence upon the probable cosmical origin of worlds. Yet such in truth has been the case. Every substance when rendered luminous gives off light of some colour or degree of refrangibility peculiar to itself, and although the eye cannot detect any difference between one character of light and another, the prism gives the means of ascertaining the quality and degree of refrangibility of the light emanating from any source however distant, and hence of gaining some knowledge of the nature of that source. If, for instance, a ray of light from a solid body in combustion is passed through a prism, a spectrum is produced which exhibits light of all colours or all degrees of refrangibility; if the light from such a body, before passing through the prism, be made to pass through gases or certain metallic vapours, the resulting spectrum is found to be crossed transversely by numbers of fine dark lines, apparently separating the various colours, or cutting the spectrum into bands. The solar spectrum is of this class; the once mysterious lines first observed by Wollaston, and subsequently by Fraunhofer, and known as "Fraunhofer's lines," have now been interpreted, chiefly by the sagacious German chemist Kirchhoff, and identified as the effects of absorption of certain of the sun's rays by chemical vapours contained in his atmosphere. The fixed stars yield spectra of the same character, but varying considerably in feature, the lines crossing the stella spectra differing in position and number from those of the sun, and one star from another, proving the stars to possess varied chemical constitutions. But there is another class of spectra, exhibited when light from other sources is passed through the prism. These consist, not of a luminous riband of light like the solar spectrum, but of bright isolated lines of coloured light with comparatively wide dark spaces separating them. Such spectra are yielded only by the light emitted from luminous gases and metals or chemical elements in the condition of incandescent vapour. Every gas or element in the state of luminous vapour yields a spectrum peculiar to itself, and no two elements when vaporized before the prism show the same combinations of luminous lines.

Now in the course of some observations upon the spectra of the fixed stars by Dr. Huggins, it occurred to that gentleman to turn his telescope, armed with a spectroscope, upon some of the brighter of the nebulæ, and great was his surprise to find that instead of yielding continuous spectra, as they must have done had their light been made up of that of a multitude of stars, they gave spectra containing only two or three isolated bright lines; such a spectrum could only be produced by some luminous gas or vapour, and of this form of matter we are now justified in declaring, upon the strength of numerous modern observations, these remarkable bodies are composed; and it is a curious and interesting fact that some of the nebulæ styled resolvable, from the fact of their exhibiting points of light like stars, yield these gaseous spectra, whence Dr. Huggins concludes that the brighter points taken for stars are in reality nuclei of greater condensation of the nebular matter: and so the fact of the apparent resolvability of a nebula affords no positive proof of its non-nebulous character.

These observations—which have been fully confirmed by Father Secchi of the Roman College—by destroying the evidence in favour of nebulæ being remote clusters, add another attestation to the probability of the truth of the nebular hypothesis, and we have now the confutation of the luminologist to add to that of the astronomers who, in the person of the illustrious Arago, asserted that the ideas of the great author of the "Mécanique Céleste" "were those only which by their grandeur, their coherence, and their mathematical character could be truly considered as forming a physical cosmogony."

Confining, then, our attention to the single object of the universe it is our task to treat of—the Moon—and without asserting as an indisputable fact that which we can never hope to know otherwise than by inference and analogy, we may assume that that body once existed in the form of a vast mass of diffused or attenuated matter, and that, by the action of gravitation upon the particles of that matter, it was condensed into a comparatively small and compact planetary body.

But while the process of condensation or compaction was going on, another important law of nature—but recently unfolded to our knowledge—was in powerful operation, the discussion of which law we reserve for a separate Chapter.

CHAPTER II.

THE GENERATION OF COSMICAL HEAT.

IN the preceding Chapter we endeavoured to show how the action of gravitation upon the particles of diffused primordial matter would result in the formation, by condensation and aggregation, of a spherical planetary body. We have now to consider another result of the gravitating action, and for this we must call to our aid a branch of scientific enquiry and investigation unrecognized as such at the period of Laplace's speculations, and which has been developed almost entirely within the past quarter of a century.

The "great philosophical doctrine of the present era of science," as the subject about to engage our attention has been justly termed, bears the title of the "Conservation of Force," or—as some ambiguity is likely to attend the definition of the term "Force"—the "Conservation of Energy." The basis of the doctrine is the broad and comprehensive natural law which teaches us that the quantity of force comprised by the universe, like the quantity of matter contained in it, is a fixed and invariable amount, which can be neither added to nor taken from, but which is for ever undergoing change and transformation from one form to another. That we cannot create force ought to be as obvious a fact as that we cannot create matter; and what we cannot create we cannot destroy. As in the universe we see no new matter created, but the same matter constantly disappearing from one form and reappearing in another, so we can find no new force ever coming into action—no description of force that is not to be referred to some previous manner of existence.

Without entering upon a metaphysical discussion of the term "force," it will be sufficient for our purpose to consider it as something which produces or resists motion, and hence we may argue that the ultimate effect of force is motion. The force of gravity on the earth results in the motion or tendency of all bodies towards its centre, and similarly, the action of gravitation upon the atoms or particles of a primeval planet resulted in the motion of those particles towards each other. We cannot conceive force otherwise than by its effects, or the motion it produces.

And force we are taught is indestructible; therefore motion must be indestructible also. But when a falling body strikes the earth, or a gunshot strikes its target, or a hammer delivers a blow upon an anvil, or a brake is pressed against a rotating wheel, motion is arrested, and it would seem natural to infer that it is destroyed. But if we say it is indestructible, what becomes of it? The philosophical answer to the question is this—that the motion of the mass becomes transferred to the particles or molecules composing it, and transformed to molecular motion, and this molecular motion manifests itself to us as heat. The particles or atoms of matter are held together by cohesion, or, in other words, by the action of molecular attraction. When heat is applied to these particles, motion is set up among them, they are set in vibration, and thus, requiring and making wider room, they urge each other apart, and the well-known *expansion by heat* is the result. If the heat be further continued a more violent molecular motion ensues, every increase of heat tending to urge the atoms further apart, till at length they overcome their cohesive attraction and move about each other, and a *liquid or molten condition* results. If the heat be still further increased, the atoms break away from their cohesive fetters altogether and leap off the mass in the form of vapour, and the matter thus assumes the *gaseous or vaporous form*. Thus we see that the phenomena of heat are phenomena of motion, and of motion only.

This mutual relation between heat and work presented itself as an embryo idea to the minds of several of the earlier philosophers, by whom it was maintained in opposition to the *material theory* which held heat to be a kind of matter or subtle fluid stored up in the inter-atomic spaces of all bodies, capable of being separated and procured from them by rubbing them together, but not generated thereby. Bacon, in his "Novum

Organum," says that "heat itself, its essence and quiddity, is motion and nothing else." Locke defines heat as "a very brisk agitation of the insensible parts of an object, which produces in us that sensation from whence we denominate the object hot; so what in our sensation is *heat*, in the object is nothing but *motion*." Descartes and his followers upheld a similar opinion. Richard Boyle, two hundred years ago, actually wrote a treatise entitled "The Mechanical Theory of Heat and Cold," and the ingenious Count Rumford made some highly interesting and significant experiments on the subject, which are described in a paper read before the Royal Society in 1798, entitled "An Inquiry concerning the Source of Heat excited by Friction." But the conceptions of these authors remained isolated and unfruitful for more than a century, and might have passed, meantime, into the oblivion of barren speculation, but for the impulse which this branch of inquiry has lately received. Now, however, they stand forth as notable instances of truth trying to force itself into recognition while yet men's minds were unprepared or disinclined to receive it. The key to the beautiful mechanical theory of heat was found by these searching minds, but the unclasping of the lock that should disclose its beauty and value was reserved for the philosophers of the present age.

Simultaneously and independently, and without even the knowledge of each other, three men, far removed from probable intercourse, conceived the same ideas and worked out nearly similar results concerning the mechanical theory of heat. Seeing that motion was convertible into heat, and heat into motion, it became of the utmost importance to determine the exact relation that existed between the two elements. The first who raised the idea to philosophic clearness was Dr. Julius Robert Mayer, a physician of Heilbronn in Germany. In certain observations connected with his medical practice it occurred to him that there must be a necessary equivalent between work and heat, a necessary numerical relation between them. "The variations of the difference of colour of arterial and venous blood directed his attention to the theory of respiration. He soon saw in the respiration of animals the origin of their motive powers, and the comparison of animals to thermic machines afterwards suggested to him the important principle with which his name will remain for ever connected,"

Next in order of publication of his results stands the name of Colding, a Danish engineer, who about the year 1843 presented a series of memoirs on the steam-engine to the Royal Society of Copenhagen, in which he put forth views almost identical with those of Mayer.

Last in publication order, but foremost in the importance of his experimental treatment of the subject, was our own countryman, Dr. Joule of Manchester. " Entirely independent of Mayer, with his mind firmly fixed upon a principle, and undismayed by the coolness with which his first labours appear to have been received, he persisted for years in his attempts to prove the invariability of the relation which subsists between heat and ordinary mechanical power." (We are quoting from Professor Tyndall's valuable work on "Heat considered as a Mode of Motion.") "He placed water in a suitable vessel, agitated the water by paddles, and determined both the amount of heat developed by the stirring of the liquid and the amount of labour expended in its production. He did the same with mercury and sperm oil. He also caused discs of cast iron to rub against each other, and measured the heat produced by their friction, and the force expended in overcoming it. He urged water through capillary tubes, and determined the amount of heat generated by the friction, of the liquid against the sides of the tubes. And the results of his experiments leave no shadow of doubt upon the mind that, under all circumstances, the quantity of heat generated by the same amount of force is fixed and invariable. A given amount of force, in causing the iron discs to rotate against each other, produced precisely the same amount of heat as when it was applied to agitate water, mercury, or sperm oil. * * * * *The absolute amount of heat* generated by the same expenditure of power, was in all cases the same."

"In this way it was found that the quantity of heat which would raise one pound of water one degree Fahrenheit in temperature, is exactly equal to what would be generated if a pound weight, after having fallen through a height of 772 feet, had its moving force destroyed by collision with the earth. Conversely, the amount of heat necessary to raise a pound of water one degree in temperature, would, if all applied mechanically, be competent to raise a pound weight 772 feet high, or it would raise 772 pounds one foot high. The term 'foot-pounds' has been introduced to express in a convenient way the lifting of one pound to the height of a foot. Thus the quantity of heat necessary to raise the temperature of a pound of water one degree Fahrenheit being taken as a standard, 772 foot-pounds constitute what is called the *mechanical equivalent* of heat."

By a process entirely different, and by an independent course of reasoning, Mayer had, a few months previous to Joule, determined this equivalent to be 771·4 foot-pounds. Such a remarkable coincidence arrived at by pursuing different routes gives this value a strong claim to accuracy, and raises the Mechanical Theory of Heat to the dignity of an exact science, and its enunciators to the foremost place in the ranks of physical philosophers.

In linking together the labours of the two remarkable men above alluded to, Prof. Tyndall remarks, that "Mayer's labours have in some measure the stamp of profound intuition, which rose however to the energy of undoubting conviction in the author's mind. Joule's labours, on the contrary, are an experimental demonstration. Mayer *thought* his theory out, and rose to its grandest applications. Joule *worked* his theory out, and gave it the solidity of natural truth. True to the speculative instinct of his country, Mayer drew large and mighty conclusions from slender premises; while the Englishman aimed above all things at the firm establishment of facts . . . . . . To each belongs a reputation which will not quickly fade, for the share he has had, not only in establishing the dynamical theory of heat, but also in leading the way towards a right appreciation of the general energies of the universe."

But from these generalities we must pass to the application of the mechanical theory of heat to our special subject. We have learnt that every form of motion is convertible into heat. We know that the falling meteor or shooting star, whose motion is impeded by friction against the earth's atmosphere, is heated thereby to a temperature of incandescence. Let us then suppose that myriads of such cosmical particles came into collision from the effect of their mutual attraction, or that the component atoms of a vast nebulous mass violently converged under the like influence. What would follow? Obviously the generation of an intense heat by the arrest of converging motion, such a heat as would result in the fusion of the whole into one mass. Mayer, in one of his most remarkable papers ("Celestial Dynamics") remarks that the "Newtonian theory of gravitation, whilst it enables us to determine, from its present form, the earth's state of aggregation in ages past, at the same time points out to us a source of heat powerful enough to produce such a state of aggregation—powerful enough to melt worlds: it teaches us to consider the molten state of a planet as the result of the mechanical union of cosmical masses, and to derive the radiation of the sun and the heat in the bowels of the earth from a common origin."

And the same laws that governed the formation of the earth, governed also the formation of the moon: the variations of Nature's operations are *quantitative* only and not *qualitative*. The Divine Will that made the earth made the moon also, and the means and mode of working were the same for both. The geological phenomena of the earth afford unmistakeable evidence of its original fluid or molten condition, and the appearance of the moon is as unmistakeably that of a body once in an igneous or molten state. The enigma of the earth's primary formation is solved by the application of the dynamical theory of heat. By this theory the generation of cosmical heat is removed from the quicksands of conjecture and established upon the firm ground of direct calculation: for the absolute amount of heat generated by the collision of a given amount of matter is (of course, with some little uncertainty) deducible from a mathematical formula. Mayer has computed the amount of heat that the matter of the earth would have generated, if it had been formed originally of only two parts drawn into collision by their mutual attraction, and has found that it would be from 0 to 32,000 or 47,000* Centigrade degrees, according as one part was infinitely small as compared with the other, or as the two parts were of equal size. Professor Helmholtz, another labourer in the same field of science, has computed the amount of heat generated by the condensation of the whole of the matter composing the solar system: this he finds would be equivalent to the heat that would be required to raise the temperature of a mass of water equal to the sum of

---

* The melting temperature of iron is 1500° Centigrade.

the masses of all the bodies of the system to 28,000,000 (twenty-eight million) degrees of the Centigrade scale.

These examples afford abundant evidence of sufficient heat having been generated by the aggregation of the matter of the moon to reduce it to a state of fusion, and so to produce, from a nebulous chaos of diffused cosmical matter, a molten body of definite outline and size.

It is requisite here to remark that fusion does not necessarily imply combustion. It has been frequently asked, How can a volcanic theory of the lunar phenomena be upheld consistently with the condition that it possesses no atmosphere to support Fire? To this we would reply that to produce a state of incandescence or a molten condition it is *not* necessary that the body be surrounded by an atmosphere. The intensely rapid motion of the particles of matter of bodies, which the dynamical theory shows to be the origin of the molten state, exists quite independently of such external matter as an atmosphere. The complex mixture of gases and vapours which we term "air," has nothing whatever to do with the fusion of substances, whatever it may have to do with their combustion. Combustion is a chemical phenomenon, due to the combination of the oxygen of that air with the heated particles of the combustible matter: oxygen is the sole supporter of combustion, and hence combustion is to be regarded rather as a phenomenon of oxygen than as a phenomenon of the matter with which that oxygen combines. The greatest intensity of heat may exist without oxygen, and consequently without combustion. In support of this argument it will be sufficient to adduce, upon the authority of Dr. Tyndall, the fact that a platinum wire can be raised to a luminous temperature and actually *fused* in a perfect vacuum.

But while the mass of condensing cosmical matter was thus accumulating and forming the globe of the moon, the heat consequent upon the aggregation of its particles was suffering some diminution from the effect of radiation. So long as the radiated heat lost fell short of the dynamical heat generated, no effect of cooling would be manifest; but when the *vis viva* of the condensing matter was all converted into its equivalent of heat, or when the accession of heat fell short of that radiated, a necessary cooling must ensue, and this cooling would be accompanied by a solidification of that part of the mass which was most free to radiate its heat into surrounding space : that part would obviously be the outer surface.

With the solidification of this external crust began the "year one" of selenological history.

The phenomena attendant upon the cooling of the mass we will consider in the next Chapter.

CHAPTER III.

THE SUBSEQUENT COOLING OF THE IGNEOUS BODY.

IN the foregoing Chapters we have endeavoured to show, by the light of modern science, first, how diffused cosmical matter was probably condensed into a planetary mass by the mutual gravitation of its particles, and secondly, how, the after destruction of the gravitative force, by the collision of the converging particles of matter, resulted in the generation of such sufficient heat as to reduce the whole mass to a molten condition. Our present task is to consider the subsequent cooling of the mass, and the phenomena attendant upon or resulting therefrom. This brief Chapter is important to our subject, as we shall have frequent occasion to refer to the leading principle we shall endeavour to illustrate in it, in subsequently treating of the causes to which the special selenological features are to be attributed.

First, then, as regards the cooling of the igneous mass that constituted the moon at the inconceivably remote period when possibly that body was really "a lesser light" shining with a luminosity of its own, due to its then incandescent state, and not simply a reflector, as it is now, of light which it receives from the sun. If we could conceive it possible that the igneous mass in the act of cooling parted with its heat from the central part first and so began to solidify from its centre, or if it had been possible for the mass to have cooled uniformly and simultaneously throughout its whole depth, or that each substratum had cooled before its superstratum, we should have had a moon whose surface would have been smooth and without any such remarkable asperities and excrescences as are now presented to our view. But these suppositions are inadmissible: on the contrary we are compelled to consider that the portion of the igneous or molten body that first cooled was its exterior surface, which, radiating its heat into surrounding space, became solid and comparatively cool while the interior retained its hot and molten condition. So that at this early stage of the moon's history it existed in the form of a solid shell inclosing a molten interior.

Now at this period of its formation, the moon's mass, partly cooled and solidified and partly molten, would be subject to the influence of two powerful molecular forces : the first of these would consist in the diminution of bulk or contraction of volume which accompanies the cooling of solidified masses of previously molten substances ; the second would arise from a phenomenon which we may here observe is by no means so generally known as from its importance it deserves to be : and as we shall have frequent occasion to refer to it as one of the chief agencies in producing the peculiar structural characteristics of the moon's surface, it may be well here to give a few examples of its action, that our reference to it hereafter may be more clearly understood.

The broad general principle of the phenomenon here referred to is this :—that fusible substances are (with a few exceptions) specifically heavier while in their molten condition than in the solidified state, or in other words, that molten matter occupies less space, weight for weight, than the same matter after it has passed from the melted to the solid condition. It follows as an obvious corollary that such substances contract in bulk in fusing or melting, and expand in becoming solid. It is this expansion upon solidification that now concerns us.

Water, as is well known, increases in density as it cools, till it reaches the temperature of 39° Fahrenheit, after which, upon a further decrease of temperature, its density begins to decrease, or in other words its bulk expands, and hence the well-known fact of ice floating in water, and the inconvenient fact of water-pipes bursting in a frost. This action in water is of the utmost importance in the grand economy of nature, and it has been accepted as a marvellous exception to the general law of substances increasing in density (or shrinking) as they decrease in temperature. Water is, however, by no means the exceptional substance that it has been so generally considered. It is a fact perfectly familiar to iron-founders, that when a mass of solid cast-iron is dropped into a pot of molten iron of identical quality, the solid is found to float persistently upon the molten metal—so persistently that when it is intentionally thrust to the bottom of the pot, it rises again the moment the submerging agency is withdrawn. As regards the amount of buoyancy we believe it may be stated in round numbers to be at least two or three per cent. It has been suggested by some who are familiar with this phenomenon that the solid mass may be kept up by a spurious buoyancy imparted to it by a film of adhering air, or that surface impurities upon the solid metal may tend to reduce the specific gravity of the mass and thereby prevent it sinking, and that the fact of floatation is not absolutely a proof of greater specific lightness. But in controversion of these suggestions, we can state as the result of experiment that pieces of cast-iron which have had their surface roughness entirely removed, leaving the bright metal exposed, still float on the molten metal, and further that when, under the influence of the great heat of the molten mass, the solid is gradually melted away, and consequently any possible surface impurities or adhering air must necessarily have been removed, the remaining portion continues to float to the last. The inevitable inference from this is that in the case of cast-iron the solid is specifically lighter than the molten, and, therefore, that in passing from the molten to the solid condition this substance undergoes expansion in bulk.

We are able to offer a confirmation of this inference in the case of cast-iron by a remarkable phenomenon well known to iron-founders, but of which we have never met with special notice. When a ladle or pot of molten iron is drawn from the melting furnace and allowed to stand at rest, the surface presents a most remarkable and suggestive appearance. Instead of remaining calm and smooth it is the scene of a lively commotion : the thin coat of scoria or molten oxide which forms on the otherwise bright surface of the metal is seen, as fast as it forms at the circumference of the pot, to be swept by active convergent currents towards the centre, where it accumulates in a patch. While this action is proceeding, the entire upper surface of the metal appears as if it were covered with animated vermicules of scoria, springing into existence at the circumference of the pot, and from thence rapidly streaming and wriggling themselves towards the centre.

Our illustration (Fig. 1) is intended, so far as such means can do so, to convey some idea of this remarkable appearance at one instant of its continued occurrence. To interpret our illustration rightly it is necessary

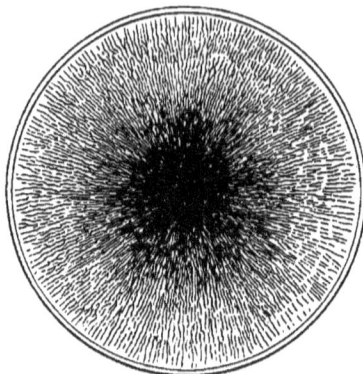

Fig. 1.

to imagine this vermicular freckling to be constantly and rapidly streaming from all points of the periphery of the pot towards the centre, where, as we have said, it accumulates in the form of a floating island. We may observe that the motion is most rapid when the hot metal is first put into the cool ladle : as the fluid metal parts with some of its heat and the ladle gets hot by absorbing it, this remarkable surface disturbance becomes less energetic.

Now if we carefully consider this peculiar action and seek a cause for the phenomenon, we shall be led to the conclusion that it arises from the expansion of that portion of the molten mass which is in contact with or close proximity to the comparatively cool sides of the ladle, which sides act as the chief agent in dispersing the heat of the melted metal. The motion of the scoria betrays that of the fluid metal beneath, and careful observation will show that the motion in question is the result of an upward current of the metal around the circumference of the ladle, as indicated by the arrows A, B, C in the accompanying sectional drawing of the ladle (Fig. 2). The upward current of the metal can actually be seen when specially looked for, at the rim of the pot, where it is deflected into the convergent horizontal direction and where it presents an eleva-

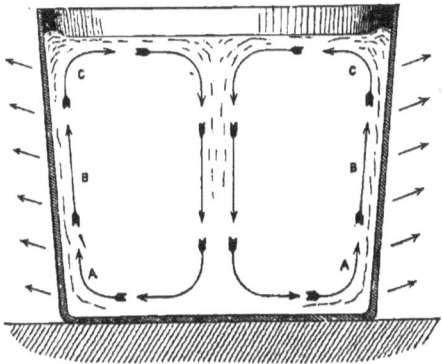

FIG. 2.

tory appearance as shown in the figure. It is difficult to assign to this effect any other cause than that of an expansion and consequent reduction of the specific gravity of the fluid metal in contact with or in close proximity to the cooler sides of the pot, as, according to the generally entertained idea that contraction universally accompanies cooling, it would be impossible for the cooler to float on the hotter metal, and the curious surface-currents above referred to would be in contrary direction to that which they invariably take, i.e., they would diverge from the centre instead of converging to it. The external arrows in the figure represent the radiation of the heat from the outer sides of the pot, which is the chief cause of cooling.

Turning from cast-iron to other metals we find further manifestations of this expansive solidification. Bismuth is a notable example. In his lectures on Heat, Dr. Tyndall exhibited an experiment in which a stout iron bottle was filled with molten bismuth, and the stopper tightly closed. The whole was set aside to cool, and as the metal within approached consolidation the bottle was rent open by its expansion, just as would have been the case had the bottle been filled with water and exposed to freezing temperature. Mercury affords another example. Thermometers which have to be exposed to Arctic temperatures are generally filled with spirit instead of quicksilver, because the latter has been found to burst the bulbs when the cold reached the congealing point of the metal, the bursting being a consequence of the expansion which accompanies the act of congelation. Silver also expands in passing from the fluid to the solid state, for we are informed by a practical refiner that solid floats on molten silver as ice floats on water; it also, as likewise do gold and copper, exhibits surface converging currents in the melting-pot like those depicted above for molten iron.

It may, however, be objected that metals are too distantly related to volcanic substances to justify inferences being drawn from their behaviour in explanation of volcanic phenomena. With a view therefore of testing the question at issue with a substance admitted as closely allied to volcanic material, we appealed to the furnace slag of iron-works. The following are extracts from the letters of an iron manufacturer of great experience * to whom we referred the question:—

"I beg to inform you that cold slag floats in molten slag in the same way cold iron floats in molten iron.

"I filled a box with hot molten slag run quickly from a blast furnace; the box was about 5½ feet square by 2 feet deep, and I dropped into the slag a piece of cold slag weighing 16 lbs., when it came to the top in a second. I pushed it down to the bottom several times and it always made its appearance at the top: indeed a small portion of it remained above the molten slag."

Here then we have a substance closely allied to volcanic material which manifests the expansile principle in question; but we may go still

* Mr. T. Heunter, Manager of the Iron-works of James Murray, Esq., of Dalmellington, Ayrshire. Another authority (Mr. Snelus, of the West Cumberland Iron Company), writes as follows: "I had a hole dug on the 'cinder-fall,' and allowed the running slag to flow through it so as to form a tolerably large pool and yet keep fluid. Any crust that formed was skimmed off. A portion of the same slag was cooled, and the solid lump thrown into the pool. It floated just at the surface." Mr. Snelus adds, by the way, that he tried "Bessemer-Pig" in the same way, and that the solid pig sunk in the molten for a minute and then *rose and floated* just at the surface, with about one twentieth of its bulk above the level of the fluid.

further and give evidence from the very fountain-head by instancing what appears to be a most cogent example of its operation which we observed on the occasion of a visit to the crater of Vesuvius in 1865 while a modified eruption was in progress. On this occasion we observed white-hot lava streaming down from apertures in the sides of a central cone within the crater and forming a lake of molten lava on the plateau or bottom of the crater; on the surface of this molten lake vast cakes of the same lava which had become solidified were floating, exactly in the same manner as ice floats in water. The solidified lava had cracked, and divided into cakes, in consequence of its contraction and also of the uprising of the accumulating fluid lava on which it floated, more and more space being thus afforded for it to separate, on account of the crater widening upwards, while through the joints or fissures the fluid lava could be seen beneath. But for the decrease in density and consequent expansion in volume which accompanied solidification, this floating of the solidified lava on the molten could not have occurred. Reference to Fig. 3,

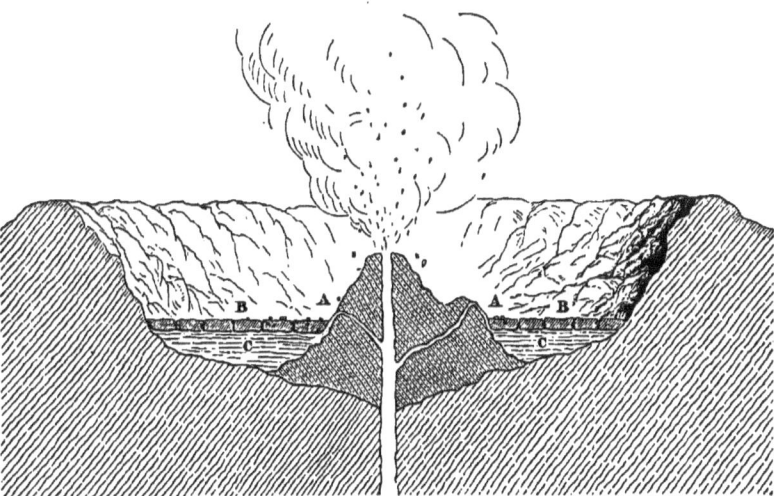

Fig. 3.

which represents a section of the crater of Vesuvius on the occasion above referred to, will perhaps assist the reader to a more clear idea of what we have endeavoured to describe. A A are the streams of white-hot lava issuing from openings in the sides of the central cone, and accumulating beneath the solidified crust B B in a lake of molten lava at C C; the solidified crust B B as it was floated upwards dividing into separate cakes as represented in Fig. 4. (See also Plate I.)

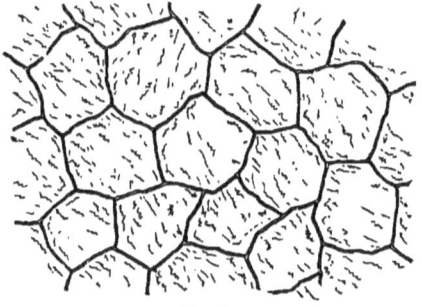

Fig. 4.

Let us now consider what would be the effect produced upon a spherical mass of molten matter in progress of cooling, first under the action of the above described expansion which precedes solidification, and then by the contraction which accompanies the cooling of a solidified body. The first portion of such a mass to part with its heat being its external surface, this portion would expand, but there being no obstacle to resist the expansion there would be no other result than a temporary slight enlargement of the sphere. This external portion would on cooling form a solid shell encompassing a more or less fluid molten nucleus, but as this interior has in its turn, on approaching the point of solidification, to expand also, and there being, so to speak, no room for its expansion, by reason of its confinement within its solid casing, what would be the consequence?—the shell would be rent or burst open, and a portion of the molten interior ejected with more or less violence according to circumstances, and many of the characteristic features of volcanic action would

SUBSEQUENT COOLING OF IGNEOUS BODY.

be thus produced: the thickness of the outer shell, the size of the vent made by the expanding matter for its escape, and other conditions conspiring to modify the nature and extent of the eruption. Thus there would result vast floodings of the exterior surface of the shell by the so extruded molten matter, volcanoes, extruded mountains, and other manifestations of eruptive phenomena. The sectional diagram (Fig. 5) will

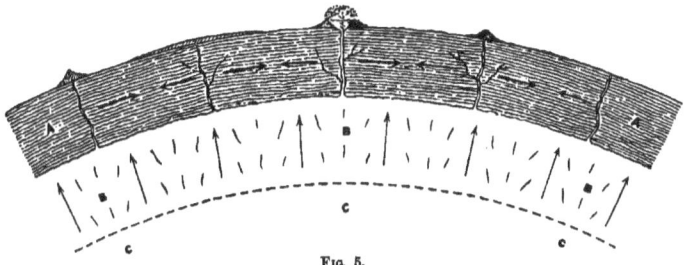

FIG. 5.

A A. The solidified crust cooling, contracting, and cracking; the cracking action enhanced by the expansion of the substratum of molten matter, B B B, which, expanding as it approaches the point of solidification, injects portions of the molten matter up through the contractile cracks, and results in producing craters, mountains of exudation, and districts flooded with extruded lava. C C C. The nucleus of intensely hot molten matter.

help to convey a clear idea of this action. Basing our reasoning on the principle we have thus enunciated, namely, that molten telluric matter expands on nearing the point of solidification, and which we have endeavoured to illustrate by reference to actual examples of its operation, we consider we are justified in assuming that such a course of volcanic phenomena has very probably occurred again and again upon the moon; that this expansion of volume which accompanies the solidification of molten matter furnishes a key to the solution of the enigma of volcanic action; and that such theories as depend upon the agency of gases, vapour, or water are at all events untenable with regard to the moon, where no gases, vapour, or water, appear to exist.

That an upheaving and ejective force has been in action with varying intensity beneath the whole of the lunar surface is manifest from the aspect of its structural details, and we are impressed with the conviction that the principle we have set forth, namely the paroxysms of expansion which successively occurred as portions of its molten interior approached solidification, supply us with a rational cause to which such vast ejective and upheaving phenomena may be assigned. Many features of terrestrial geology likewise require such an expansive force whereby to explain them; we therefore venture to recommend this source and cause of ejective action to the careful consideration of geologists.

When the molten substratum had burst its confines, ejected its superfluous matter, and produced the resulting volcanic features, it would, after final solidification, resume the normal process of contraction upon cooling, and so retreat or shrink away from the external shell. Let us now consider what would be the result of this. Evidently the external shell or crust would become relatively too large to remain at all points in close contact with the subjacent matter. The consequence of too large a solid shell having to accommodate itself to a shrunken body underneath, is that the skin, so to term the outer stratum of solid matter, becomes shrivelled up into alternate ridges and depressions, or wrinkles. In its attempt to crush down and follow the contracting substratum, it would have to displace the superabundant or superfluous material of its former larger surface by thrusting it (by the action of tangential force) into

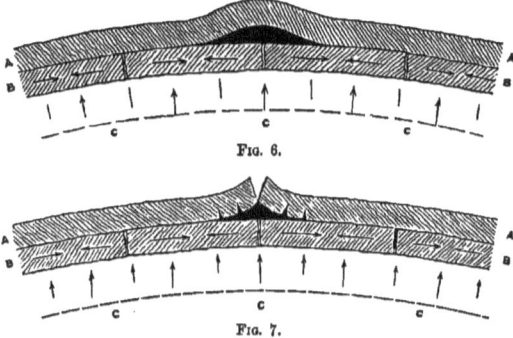

Fig. 6.

Fig. 7.

undulating ridges as in Fig. 6, or broken elevated ridges as in Fig. 7, or overlappings of the outer crust as in Fig. 8, or ridges capped by more or less fluid molten matter extruded from beneath, as indicated in Fig. 9, a class of action which might occur contemporaneously with the elevation of the ridge or subsequently to its formation.

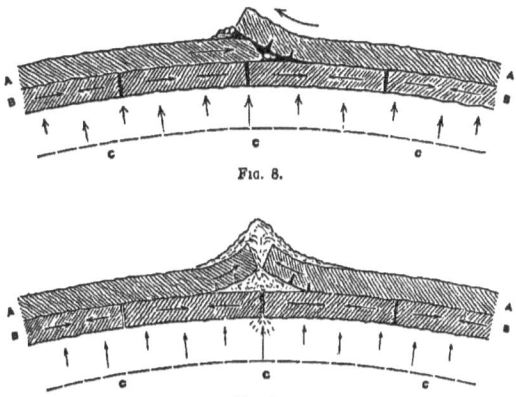

Fig. 8.

Fig. 9.

A long-kept shrivelled apple affords an apt illustration of this wrinkle theory; another example may be observed in the human face and hand, when age has caused the flesh to shrink and so leave the comparatively unshrinking skin relatively too large as a covering for it. We illustrate both of these examples by actual photographs of the respective objects, which are reproduced on Plate II. Whenever an outer covering has to accommodate and apply itself to an interior body that has become too small for it, wrinkles are inevitably produced. The same action that shrivels the human skin into creases and wrinkles, has also shrivelled certain regions of the igneous crust of the earth. A map of a mountainous part of our globe affords abundant evidence of such a cause having been in action; such maps are pictures of wrinkles. Several parts of the lunar surface, as we shall by-and-by see, present us with the same appearances in a modified degree.

To the few primary causes we have set forth in this chapter—to the alternate expansion and contraction of successive strata of the lunar sphere, when in a state of transition from an igneous and molten to a cooled and solidified condition, we believe we shall be able to refer well nigh all the remarkable and characteristic features of the lunar surface which will come under our notice in the course of our survey.

CHAPTER IV.

THE FORM, MAGNITUDE, WEIGHT, AND DENSITY OF THE LUNAR GLOBE.

WE have not hitherto had occasion to refer to what we may term the physical elements of the moon: by which we mean the various data concerning form, size, weight, density, &c. of that body, derived from observation and calculation. To this purpose, therefore, we will now devote a few pages, confining ourselves to such matters as specially bear upon the requirements of our subject, omitting such as are irrelevant to our purpose, and touching but lightly upon such as are commonly known, or are explained in ordinary elementary treatises on astronomy.

First, then, as regards the form of the moon. The form of the lunar disc, when fully illuminated, we perceive to be a perfect circle; that is to say, the measured diameters in all directions are equal; and we are therefore led to infer that the real form of the moon is that of a perfect sphere. We know that the earth and the rest of the planets of our system are spheroidal, or more or less flattened at the poles, and we also know that this flattening is a consequence of axial rotation; the extent of the flattening, or the oblateness of the spheroid, depending upon the speed of that rotation. But in the case of the moon the axial rotation is so slow that the flattening produced thereby, although it must exist, is so slight as to be imperceptible to our observation. We might therefore conclude that the moon is a perfectly spherical body, did not theory step in to show us that there is another cause by which its form is disturbed. Assuming the moon to have been once in a fluid state, it is demonstrable that the attraction of the earth would accumulate a mass of matter, like a tidal elevation, in the direction of a line joining the centres of the two bodies: and as a consequence, the real shape of the moon must be an ellipsoid, or somewhat egg-shaped body, the major axis of which is directed towards the earth. That some such phenomenon has obtained is evident from the coincidence of the times of orbital revolution and axial rotation of the lunar sphere. "It would be against all probability," says Laplace, "to suppose that these two motions had been at their origin perfectly equal;" but it is sufficient that their primitive difference was but small, in which case the constant attraction by the earth of the protruberant part of the moon would establish the equality which at present exists.

It is, however, sufficient for all purposes with which we are concerned to regard the moon as a sphere, and the next point to be considered is its size. To determine this, two data are necessary—its apparent or angular diameter, and its distance from the earth. The first of these is obtained by measuring the angle comprised between two lines directed from the eye to two opposite "limbs" or edges of the moon. If, for instance, we were to take a pair of compasses and, placing the joint at the eye, open out the legs till the two points appear to touch two opposite edges of the moon, the two legs would be inclined at an angle which would represent the diameter of the moon, and this angle we could measure by applying a divided arc or protractor to the compasses. In practice this measurement is made by means of telescopes attached to accurately divided circles; the difference between the readings of the circle when the telescope is directed to opposite limbs of the moon giving its angular diameter at the time of the observation. But from the fact that the orbit of the moon is an ellipse, it is evident that she is at some times much nearer to us than at others, and, as a consequence, her apparent magnitude is variable: there is also a slight variation depending upon the altitude of the moon at the time of the measure; the mean diameter, however, or the diameter at mean distance from the centre of the earth has, from long course of observation, been found to be 31′ 9″.

To convert this apparent angular diameter into real linear measurement, it is necessary to know either the distance of the moon from the earth, or in astronomical language as leading to a knowledge of that distance, what is the amount of the moon's *parallax*. Parallax, generally, is an apparent change of position of an object arising from change of the point of view. The parallax of a heavenly body is the angle which the earth would subtend if it were seen from that body. Supposing an observer on the moon could measure the earth's angular diameter, just as we measure that of the moon, his measurement would represent what is called the parallax of the moon. But we cannot go to the moon to make such a measurement; nevertheless there is a simple method, explained in most treatises on astronomy, which consists in observing the moon from stations on the earth widely separated, and by which we can obtain a precisely similar result. Without detailing the process, it is sufficient for us to know that the angle which would be subtended by the earth if seen from the moon, or the moon's parallax, is according to the latest determination, equal to 1° 54′ 5″. This value, however, varies considerably with the variations of distance due to the elliptic orbit of the moon: the number we have given represents the mean parallax, or the parallax at mean distance.

But we have to turn these angular measurements into miles. To effect this we have only to work a simple rule of three sum. It will easily be understood that, as the angular diameter of the earth seen from the moon is to the angular diameter of the moon seen from the earth, so is the diameter of the earth in miles to the diameter of the moon in miles. The diameter of the earth we know to be 7912 miles: putting this therefore in its proper place in the proportion sum, and duly working it out by the schoolboy's rule, we get:—

$$1° . 54' . 5'' \; : \; 31' . 9'' \; :: \; 7912 \text{ miles} \; : \; 2160 \text{ miles}.$$

And 2160 miles is therefore the diameter of the lunar globe.

Knowing the diameter, we can easily obtain the other elements of magnitude. According to the well-known relation of the diameter of a sphere to its area, we find the area of the moon to be 14,657,000 square miles: or half that number, 7,328,500 miles, as the area of the hemisphere at any one time presented to our view. And similarly, from the relation of the solidity of a sphere to its diameter, we find the solid contents of the moon to be 5276 millions of cubic miles of matter.

Comparing these data with corresponding dimensions of the earth, we find that the diameter of the moon is $\frac{1}{3\cdot 663}$; the area $\frac{1}{13\cdot 4243}$; and the volume $\frac{1}{49\cdot 1805}$, of the respective elements of the earth. Those who prefer a graphical to a numerical comparison, may judge of the sizes of the two bodies by the accompanying illustration (Fig. 10).

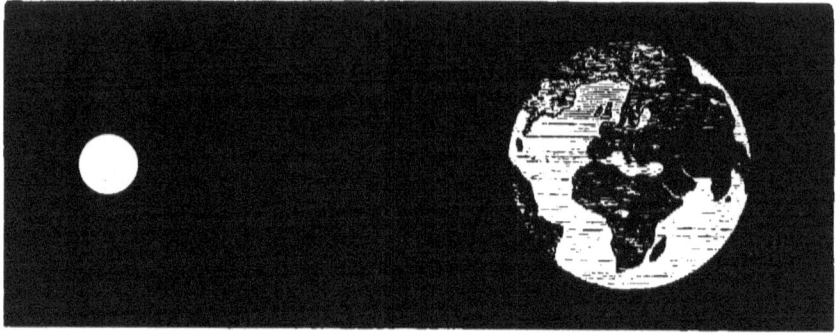

FIG. 10.

To gain an idea of their distance from each other it is necessary to suppose the two discs in the diagram to be five feet apart; the real distance of the moon from the earth being about 238,790 miles at its mean position.

Next, we come to what is technically termed the *mass*, but what in common language we may call the *weight* of the moon. It is important to know this, because the weight of a body taken in connection with its size furnishes us with a knowledge of its density, or the specific gravity of the material of which it is composed. But it is not quite so easy to determine the mass as the dimensions of the moon: to *measure* it, we have seen is easy enough; to *weigh* it is a comparatively difficult matter. To solve the problem we have to appeal to Newton's law of universal gravitation. This law teaches us that every particle of matter in the universe attracts every other particle with a force which is *directly proportional to the mass*, and inversely proportional to the square of the distance of the attracting particles. There are several methods by which this law is applied to the measurement of the mass of the moon. One of the simplest is by the agency of the Tides. We know that the moon, attracting the waters, produces a certain amount of elevation of the aqueous covering of the earth; and we know that the sun produces also a like elevation, but to a much smaller extent, by reason of its much greater distance. Now measuring accurately the heights of the solar and lunar tides, and making allowance for the difference of distance of the sun and moon from the earth, we can compare directly the effect that is due to the sun with the effect that is due to the moon: and since the masses of the two bodies are just in proportion to the effects they produce, it is evident that we have a comparison between the mass of the sun and that of the moon; and knowing what is the sun's mass we can, by simple proportion, find that of the moon. Another method is as follows:—The moon is retained in her orbital path by the attraction of the earth; if it were not for this attraction she would fly off from her course in a tangential line. She has thus a constant tendency to quit her orbit, which the earth's attraction as constantly overcomes. It is evident from this that the earth pulls the moon towards itself by a definite amount in every second of time. But while the earth is pulling the moon, the moon is also pulling the earth: they are pulling each other together; and moreover each is exerting a pull which is *proportional to its mass*. Knowing, then, the mass of the earth, which we do with considerable accuracy, we can find what share of the whole pulling force is due to it, the residue being the moon's share: the proportion which this residue bears to the earth's share gives us the proportion of the moon's mass to that of the earth, and hence the mass of the moon.

There are yet two other methods: one depending upon the phenomena of nutation, or the attraction of the sun and moon upon the protruberant matter of the terrestrial spheroid; and the other upon a displacement of the centre of gravity of the earth and moon, which shows itself in observations of the sun. By each and all of these methods has the lunar mass been at various times determined, and it has been found, as the latest and best accepted value, that the mass of the moon is *one-eightieth* that of the earth.

From the known diameter of the earth we ascertain that its volume is 259,360 millions of cubic miles: and from the various experiments that have been made to determine the mean density of the earth, it has been found that that mean density is about $5\frac{1}{2}$ times that of water; that is to say, the earth weighs $5\frac{1}{2}$ times heavier than would a sphere of water of equal size. Now a cubic foot of water weighs 62·3211 pounds, and from this we can find by simple multiplication what is the weight of a cubic mile of water, and, similarly, what would be the weight of 259,360 cubic miles of water, and the last result multiplied by $5\frac{1}{2}$ will give the weight of the earth in tons: The calculation, although extremely simple, involves a confusing heap of figures; but the result, which is all that concerns us, is, that the weight of the earth is 5842 trillions of tons: and since, as we have above stated, the mass of the earth is 80 times that of the moon, it follows that the weight of the moon is 73 trillions of tons.

The cubical contents of a body compared with its weight gives us its density. In the moon we have 5276 millions of cubic miles of matter, the total weight of which is 73 trillions of tons. Now, 5276 millions of cubic miles of water would weigh about $21\frac{1}{2}$ trillions of tons; and as this number is to 73 as 1 is to 3·4, it is clear that the density of the lunar matter is 3·4 greater than water: and inasmuch as the earth is $5\frac{1}{2}$ times denser than water, we see that the moon is about 0·62 as dense as the earth, or that the material of the moon is lighter, bulk for bulk, than the mean material of the terraqueous globe in the proportion of 62 to 100, or, nearly, 6 to 10. This specific gravity of the lunar material (3·4) we may remark is about the same as that of flint glass or the diamond: and curiously enough it nearly coincides with that of some of the aërolites that have from time to time fallen to the earth; hence support has been claimed for the theory that these bodies were originally fragments of lunar matter, probably ejected at some time from the lunar volcanoes with such force as to propel them so far within the sphere of the earth's attraction that they have ultimately been drawn to its surface.

Reverting, now, to the mass of the moon: we must bear in mind that the mass or weight of a planetary body determines the weight of all objects on its surface. What we call a pound on the earth, would not be a pound on the moon; for the following reason:—When we say that such and such an object weighs so much, we really mean that it is attracted towards the earth with a certain force depending upon its own weight. This attraction we call gravity; and the falling of a weight to the earth is an example of the action of the law of universal gravitation. The earth and the weight fall together—or are held together if the weight is in contact with the earth—with a force which depends directly upon the mass of the two, and upon the distance between them. Newton proved that the attraction of a sphere upon external objects is precisely as if the whole of its matter were contained at its centre. So that the attractive force of the earth upon a ton weight at its surface, is the attraction which 5842 trillions of tons exert upon one ton situated 3956 miles (the radius of the earth) distant. If the weight of the earth were only half the above quantity, it is clear that the attraction would be only half what it is; and hence the ton weight, being pulled by only half the force, would only be equal to half a ton; that is to say, only half as much muscular force (or any other force but gravity) would be required to lift it. It is plain, therefore, that what weighs a pound on the earth could not weigh a pound on the moon, which is only $\frac{1}{80}$ of the weight of the earth. What, then, is the relation between a pound on the earth and the same mass of matter on the moon? It would seem, since the moon's mass is $\frac{1}{80}$ of the earth, that the pound transported to the moon ought to weigh the eightieth part of a pound there; and so it would if the distance from the centre of the moon to its surface were the same as the distance of the centre of the earth from its surface. But the radius of the moon is only $\frac{1}{3.665}$ that of the earth; and the force of gravity varies *inversely as the square of the distance* between the centres of the gravitating masses. So that the attraction by the moon of a body at its surface, as compared with that of the earth, is $\frac{1}{80}$ divided by the square of $\frac{1}{3.665}$; and this, worked out, is equal to $\frac{1}{6}$. The force of gravity upon the moon is, therefore, $\frac{1}{6}$ of that on the earth; and hence a pound upon the earth would be little more than $2\frac{1}{2}$ ounces on the moon; and it follows as a consequence that any force, such as muscular exertion, or the energy of chemical, plutonic or explosive forces, would be six times more effective upon the moon than upon the earth. A man who could jump six feet from the earth, could with the same muscular effort jump thirty-six feet from the moon; the explosive energy that would project a body a mile above the earth would project a like body six miles above the surface of the moon.

It is the practice, in elementary and popular treatises on astronomy, to state merely the numerical results in giving data such as those embodied in the foregoing pages; and uninitiated readers, not knowing the means by which the figures are arrived at, are sometimes disposed to regard them with a certain amount of doubt or uncertainty. On this account we have thought it advisable to give, in as brief and concise a form as possible, the various steps by which these seemingly unattainable results are obtained.

The data explained in the foregoing text are here collected to facilitate reference.

Diameter of Moon .	2160 miles	$\frac{1}{3685}$ that of earth.
Area	14,657,000 square miles . .	$\frac{1}{15496}$,, ,,
Area of the visible hemisphere	7,328,500 square miles	
Solid contents . . .	5276 millions of cubic miles .	$\frac{1}{49158}$,, ,,
Mass	73 trillions of tons . . .	$\frac{1}{80}$,, ,,
Density	3·39 (water = 1) . .	0·62 ,, ,,
Force of gravity at surface	. .	$\frac{1}{6}$,, ,,
Mean distance from earth	. 238,700 miles.	

CHAPTER V.

ON THE EXISTENCE OR NON-EXISTENCE OF A LUNAR ATMOSPHERE.

AT the close of the preceding chapter we stated that any force acting in opposition to that of gravity would be six times more effective on the moon than on the earth. But, in fact, it would in many cases be still more so ; at all events, so far as projectile forces are concerned ; for the reason that "the powerful coercer of projectile range," as the earth's atmosphere has been termed, has no counterpart, or at most a very disproportionate one, upon the moon.

The existence of an atmosphere surrounding the moon has been the subject of considerable controversy, and a great deal of evidence on both sides of the question has been offered from time to time, and is to be found scattered through the records of various classes of observations. Some of the more important items of this evidence it is our purpose to set forth in the course of the present chapter.

With the phenomena of the terrestrial atmosphere, with the effects that are attributable to it, we are all well familiar, and our best course therefore is to examine, as far as we are able, whether counterparts of any of these effects are manifested upon the moon. For instance, the clouds that are generated in and float through our air would, to an observer on the moon, appear as ever changing bright or dusky spots, obliterating certain of the permanent details of the earth's surface, and probably skirting the terrestrial disc, like the changing belts we perceive on the planet Jupiter, or diversifying its features with less regularity, after the manner exhibited by the planet Mars. If such clouds existed on the moon it is evident that the details of its surface must be, from time to time, similarly obscured ; but no trace of such obscuration has ever been detected. When the moon is observed with high telescopic powers, all its details come out sharp and clear, without the least appearance of change or the slightest symptoms of cloudiness other than the occasional want of general definition, which may be proved to be the result of unsteadiness or want of homogeneity in our own atmosphere; for we must tell the uninitiated that nights of pure, good definition, such as give the astronomer opportunity of examining with high powers the minute details of planetary features, are very few and far between. Out of the three hundred and sixty-five nights of a year there are probably not a dozen that an astronomer can call really fine: usually, even on nights that are to all common appearance superbly brilliant, some strata of air of different densities or temperatures, or in rapid motion, intervene between the observer and the object of his observation, and through these, owing to the ever-changing refractions which the rays of light coming from the object suffer in their course, observation of the delicate markings of a planet is impossible: all is blurred and confused, and nothing but bolder features can be recognized. It has in consequence sometimes happened that a slight indistinctness of some minute detail of the moon has been attributed to clouds or mists at the lunar surface, whereas the real cause has been only a bad condition of our own atmosphere. It may be confidently asserted that when all indistinctness due to terrestrial causes is taken account of or eliminated, there remain no traces whatever of any clouds or mists upon the surface of the moon.

This is but one proof against the existence of a lunar atmosphere, and, it may be argued, not a very conclusive one; because there may still be an atmosphere, though it be not sufficiently aqueous to condense into clouds and not sufficiently dense to obscure the lunar details. The probable existence of an atmosphere of such a character used to be inferred from a phenomenon seen during total eclipses of the sun. On these occasions the black body of the moon is invariably surrounded by a luminous halo, or glory, to which the name "corona" has been applied; and, further, besides this corona, apparently floating in it and sometimes seemingly attached to the black edge of the moon, are seen masses of cloudlike matter of a bright red colour, which, from the form in which they were first seen and from their flame-like tinge, have become universally known as the "red-flames." It used to be said that this corona could only be the consequence of a lunar atmosphere lit up as it were by the sun's rays shining through it, after the manner of a sunbeam lighting up the atmosphere of a dusty chamber; and the red flames were held by those who first observed them to be clouds of denser matter floating in the said atmosphere, and refracting the red rays of solar light as our own clouds are seen to do at sunrise and sunset. But the evidence obtained, both by simple telescopic observation and by the spectroscope, from recent extensively observed eclipses of the sun has set this question quite at rest; for it has been settled finally and indisputably that both the above appearances pertain to the sun, and have nothing whatever to do with the moon.

The occurrence of a solar eclipse offers other means in addition to the foregoing whereby a lunar atmosphere would be detected. We know that all gases and vapours absorb some portion of any light which may shine through them. If then our satellite had an atmosphere, its black nucleus when seen projected against the bright sun in an eclipse would be surrounded by a sort of penumbra, or zone of shadow, in contact with its edge, somewhat like that we have shown in an exaggerated degree in the annexed cut (Fig. 11), and the passage of this penumbra over solar spots

Fig. 11.

and other features of the solar photosphere would to some extent obscure the more minute details of such features. No such dusky band has however been at any time observed. On the contrary, a band somewhat brighter than the general surface of the sun has frequently been seen in contact with the black edge of the moon: this in its turn was held to indicate an atmosphere about the moon; but Sir George Airy has shown that a lunar atmosphere, if it really did exist, could not produce such an appearance, and that the cause of it must be sought in other directions. If this effect were really due to the passage of the solar rays through a lunar atmosphere a similar effect ought to be produced by the passage of the sun's rays through the terrestrial atmosphere : and we might hence expect to see the shadow of the earth projected on the moon during a lunar eclipse surrounded by a sort of bright zone or halo: we need hardly say such an appearance has never manifested itself. Similarly as we stated that the delicate details of solar spots would be obscured by a lunar atmosphere, small stars passing behind the moon would suffer some diminution in brightness as they approached apparent contact with the moon's edge : this fading has been watched for on many occasions, and in a few cases such an appearance has been suspected, but in by far the majority of instances nothing like a diminution of brightness or change of colour of the stars has been seen; stars of the smallest magnitude visible under such circumstances retain their feeble lustre unimpaired up to the moment of their disappearance behind the moon's limb.

Again, in a solar eclipse, even if there were an atmosphere about the moon not sufficiently dense to form a hazy outline or impair the distinctness of the details of a solar spot, it would still manifest its existence in another way. As the moon advances upon the sun's disc the latter assumes, of course, a crescent form. Now if air or vapour enveloped the moon, the exceedingly delicate cusps of this crescent would be distorted or turned out of shape. Instead of remaining symmetrical, like the lower one in the annexed drawing (Fig. 12), they would be bent or deformed after the manner we have shown in the upper one. The slightest symptom of a distortion like this could not fail to obtrude itself upon an observer's eye; but in no instance has anything of the kind been seen.

Reverting to the consequences of the terrestrial atmosphere : one of the most striking of these is the phenomenon of diffused daylight, which we need hardly remind the reader is produced by the scattering or diffusion of the sun's rays among the minute particles of vapour composing or contained in that atmosphere. Were it not for this reflexion and diffusion of the sun's light, those parts of our earth not exposed to direct sunshine would be hidden in darkness, receiving no illumination beyond the feeble amount that might be reflected from proximate terrestrial objects

Fig. 12.

actually illuminated by direct sunlight. Twilight is a consequence of this reflexion of light by the atmosphere when the sun is below the horizon. If, then, an atmosphere enveloped the moon, we should see by diffused light those parts of the lunar details that are not receiving the direct solar beams; and before the sun rose and after it had set upon any region of the moon, that region would still be partially illuminated by a twilight. But, on the contrary, the shadowed portions of a lunar landscape are pitchy black, without a trace of diffused-light illumination, and the effects that a twilight would produce are entirely absent from the moon. Once, indeed, one observer, Schroeter, noticed something which he suspected was due to an effect of this kind: when the moon exhibited itself as a very slender crescent, he discovered a faint crepuscular light, extending from each of the cusps along the circumference of the unenlightened part of the disc, and he inferred from estimates of the length and breadth of the line of light that there was an atmosphere about the moon of 5376 feet in height. This is the only instance on record, we believe, of such an appearance being seen.

Spectrum analysis would also betray the existence of a lunar atmosphere. The solar rays falling on the moon are reflected from its surface to the earth. If, then, an atmosphere existed, it is plain that the solar rays must first pass through such atmosphere to reach the reflecting surface, and returning from thence, again pass through it on their way to the earth; so that they must in reality pass through virtually twice the thickness of any atmosphere that may cover the moon. And if there be any such atmosphere, the spectrum formed by the moon's light, that is, by the sun's light reflected from the moon, would be modified in such a manner as to exhibit absorption-lines different from those found in the spectrum of the direct solar rays, just as the absorption-lines vary according as the sun's rays have to pass through a thinner or a denser stratum of the terrestrial atmosphere. Guided by this reasoning, Drs. Huggins and Miller made numerous observations upon the spectrum of the moon's light, which are detailed in the "Philosophical Transactions" for the year 1864; and their result, quoting the words of the report, was "that the spectrum analysis of the light reflected from the moon is wholly negative as to the existence of any considerable lunar atmosphere."

Upon another occasion, Dr. Huggins made an analogous observation of the spectrum of a star at the moment of its occultation, which observation he records in the following words :—" When an observation is made of the spectrum of a star a little before, or at the moment of its occultation by the dark limb of the moon, several phenomena characteristic of the passage of the star's light through an atmosphere might possibly present themselves to the observer. If a lunar atmosphere exist, which either by the substances of which it is composed, or by the vapours diffused through it, can exert a selective absorption upon the star's light, this absorption would be indicated to us by the appearance in the spectrum of new dark lines immediately before the star is occulted by the moon."

"If finely divided matter, aqueous or otherwise, were present about the moon, the red rays of the star's light would be enfeebled in a smaller degree than the rays of higher refrangibilities."

"If there be about the moon an atmosphere free from vapour, and possessing no absorptive power, but of some density, then the spectrum would not be extinguished by the moon's limb at the same instant throughout its length. The violet and blue rays would lie behind the red rays."

"I carefully observed the disappearance of the spectrum of ε Piscium at its occultation of January 4, 1865, for these phenomena; but no signs of a lunar atmosphere were detected."

But perhaps the strongest evidence of the non-existence of any appreciable lunar atmosphere is afforded by the non-refraction of the light of a star passing behind the edge of the lunar disc. Refraction, we know, is a bending of the rays of light coming from any object, caused by their passage through strata of transparent matter of different densities; we have a familiar example in the apparent bending of a stick when half plunged into water. There is a simple schoolboy's experiment which illustrates refraction in a very cogent manner, but which we should, from its very simplicity, hesitate to recall to the reader's mind did it not very aptly represent the actual case we wish to exemplify. A coin is placed on the bottom of an empty basin, and the eye is brought into such a position that the coin is just hidden behind the basin's rim. Water is then poured into the basin and, without the eye being moved from its former place, as the depth of water increases, the coin is brought by degrees fully into view; the water refracting or turning out of their course the rays of light coming from the coin, and lifting them, as it were, over the edge of the basin. Now a perfectly similar phenomenon takes place at every sunrise and sunset on the earth. When the sun is really below the horizon, it is nevertheless still visible to us because it is *brought up* by the refraction of its light by the dense stratum of atmosphere through which the rays have to pass. The sun is, therefore, exactly represented by the coin at the bottom of the basin in the boy's experiment, the atmosphere answers to the water, and the horizon to the rim or edge of the basin. If there were no atmosphere about the earth, the sun would not be so brought up above the horizon, and, as a consequence, it would set earlier and rise later by about a minute than it really does. This, of course, applies not merely to the sun, but to all celestial bodies that rise and set. Every planet and every star remains a shorter time below the horizon than it would if there were no atmosphere surrounding the earth.

To apply this to the point we are discussing. The moon in her orbital course across the heavens is continually passing before, or occulting, some of the stars that so thickly stud her apparent path. And when we see a star thus pass behind the lunar disc on one side and come out again on the other side, we are virtually observing the setting and rising of that star upon the moon. If, then, the moon had an atmosphere, it is clear, from analogy to the case of the earth, that the star must disappear later and reappear sooner than if it has no atmosphere : just as a star remains too short a time below the earth's horizon, or behind the earth, in consequence of the terrestrial atmosphere, so would a star remain too short a time behind the moon if an atmosphere surrounded that body. The point is settled in this way :—The moon's apparent diameter has been measured over and over again and is known with great accuracy ; the rate of her motion across the sky is also known with perfect accuracy : hence it is easy to calculate how long the moon will take to travel across a part of the sky exactly equal in length to her own diameter. Supposing, then, that we observe a star pass behind the moon and out again, it is clear that, if there be no atmosphere, the interval of time during which it remains occulted ought to be exactly equal to the computed time which the moon would take to pass over the star. If, however, from the existence of a lunar atmosphere, the star disappears too late and reappears too soon, as we have seen it would, these two intervals will not agree; the computed time will be greater than the observed time, and the difference, if any there be, will represent the amount of refraction the star's light has sustained or suffered, and hence the extent of atmosphere it has had to pass through.

Comparisons of these two intervals of time have been repeatedly made, the most recent and most extensive was executed under the direction of the Astronomer-Royal several years ago, and it was based upon no less than 296 occultation observations. In this determination the measured or telescopic semidiameter of the moon was compared with the semidiameter deduced from the occultations, upon the above principle, and it was found that the telescopic semidiameter was greater than the occultation semidiameter by two seconds of angular measurement or by about a thousandth part of the whole diameter of the moon. Sir George Airy, commenting on this result, says that it appears to him that the origin of this difference is to be sought in one of two causes. " Either it is due to irradiation* of the telescopic semidiameter, and I do not doubt that a part at least of the two seconds is to be ascribed to that cause; or it may be due to refraction by the moon's atmosphere. If the whole two seconds were caused by atmospheric refraction this would imply a horizontal refraction of one second, which is only $\frac{1}{2000}$ part of the earth's horizontal refraction. It is possible that an atmosphere competent to produce this refraction would not make itself visible in any other way." This result accords well, considering the relative accuracy of the means employed, with that obtained a century ago by the French astronomer Du Séjour, who made a rigorous examination of the subject founded on observations of the solar eclipse of 1764. He concluded that the horizontal refraction produced by a possible lunar atmosphere amounted to $1''\cdot5$—a second and a half—or about $\frac{1}{1400}$ of that produced by the earth's atmosphere. The greater weight is of course to be allowed to the more recent determination in consideration of the large number of accurate observations upon which it was based.

But an atmosphere 2,000 times rarer than our air can scarcely be regarded as an atmosphere at all. The contents of an air-pump receiver can seldom be rarefied to a greater extent than to about $\frac{1}{1000}$ of the density of air at the earth's surface, with the best of pneumatic machines; and the lunar atmosphere, if it exist at all, is thus proved to be twice as attenuated as what we are accustomed to recognise as a vacuum. In discussing the physical phenomena of the lunar surface, we are, therefore, perfectly justified in omitting all considerations of an atmosphere, and adapting our arguments to the non-existence of such an appendage.

And if there be no air upon the moon, we are almost forced to conclude that there can be no water; for if water covered any part of the lunar globe it must be vaporised under the influence of the long period of uninterrupted sunshine (upwards of 300 hours) that constitutes the lunar day, and would manifest itself in the form of clouds or mists obscuring certain parts of the surface. But, as we have already said, no such

* Irradiation is an ocular phenomenon in virtue of which all strongly illuminated objects appear to the eye to be larger than they really are. The impression produced by light upon the retina appears to extend itself around the focal image formed by the lenses of the eye. It is from the effect of irradiation that a white disc on a black ground looks larger than a black disc of the same size on a white ground.

obliteration of details ever takes place; and, as we have further seen, no evidence of aqueous vapour is manifested upon the occasion of spectrum observations. Since, then, the effects of watery vapour are absent, we are forced to conclude that the cause is absent also.

Those parts of the moon which the ancient astronomers assumed, from their comparatively smooth and dusky appearance, to be seas, have long since been discovered to be merely extensive regions of less reflective surface material; for the telescope reveals to us irregularities and asperities covering well nigh the whole of them, which asperities could not be seen if they were covered with water; unless, indeed, we admit the possibility of seeing to the bottom of the water, not only perpendicularly, but obliquely. Some observers have noticed features that have led them to suppose that water was at one time present upon the moon, and has left its traces in the form of appearances of erosive action in some parts. But if water ever existed, where is it now? One writer, it is true, has suggested as possible, that whatever air, and we presume he would include whatever water also, the moon may possess, is hidden away in sublunarean caves and hollows; but even if water existed in these places it must sometimes assume the vapoury form, and thus make its presence known.

Sir John Herschel pointed out that if any moisture exists upon the moon, it must be in a continual state of migration from the illuminated or hot, to the unilluminated or cold side of the lunar globe. The alternations of temperature, from the heat produced by the unmitigated sunshine of 14 days' duration, to the intensity of cold resulting from the absence of any sunshine whatever for an equal period, must, he argued, produce an action similar to that of the *cryophorus* in transporting the lunar moisture from one hemisphere to the other. The cryophorus is a little instrument invented by the late Dr. Wollaston; it consists of two bulbs of glass connected by a bent tube, in the manner shown in the annexed illustration, fig. 13. One of the bulbs, A, is half-filled with water, and, all air being exhausted, the instrument is hermetically sealed, leaving nothing within but the water and the aqueous vapour which rises therefrom in the absence of atmospheric pressure. When the empty bulb, B, is placed in a freezing mixture, a rapid condensation of this vapour

NON-EXISTENCE OF A LUNAR ATMOSPHERE.

takes place within it, and as a consequence the water in the bulb A gives off more vapour. The abstraction of heat from the water, which is a

FIG. 13.

natural consequence of this evaporation, causes it to freeze into a solid mass of ice. Now upon the moon the same phenomenon would occur did the material exist there to supply it. In the accompanying diagram let A represent the illuminated or heated hemisphere of the moon, and B

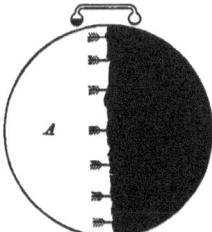

FIG. 14.

the dark or cold hemisphere; the former being probably at a temperature of 300° above, and the latter 200° below Fahrenheit's zero. Upon the above principle, if moisture existed upon A it would become vaporised, and the vapour would migrate over to B, and deposit itself there as hoar-frost; it would, therefore, manifest itself to us while in the act of migrating by clouding or dimming the details about the boundary of the illuminated hemisphere. The sun, rising upon any point upon the margin of the dark hemisphere, would have to shine through a bed of moisture, and we may justly suppose, if this were the case, that the tops of mountains catching the first beams of sunlight would be tinged with colour, or be lit up at first with but a faint illumination, just as we see in the case of terrestrial mountains whose summits catch the first, or receive the last beams of the rising or setting sun. Nothing of this kind is, however, perceptible: when the solar rays tip the lofty peaks of lunar mountains, these shine at once with brilliant light, quite as vivid as any of those parts that receive less horizontal illumination, or upon which the sun is almost perpendicularly shining.

All the evidence, then, that we have the means of obtaining, goes to prove that neither air nor water exists upon the moon. Two complicating elements affecting all questions relating to the geology of the terraqueous globe we inhabit may thus be dismissed from our minds while considering the physical features of the lunar surface. Fire on the one hand and water on the other, are the agents to which the configurations of the earth's surface are referable: the first of these produced the igneous rocks that form the veritable foundations of the earth, the second has given rise to the superstructure of deposits that constitute the secondary and tertiary formations: were these last removed from the surface of our planet, so as to lay bare its original igneous crust, that crust, so far as reasoning can picture it to us, would probably not differ essentially from the visible surface of the moon. In considering the causes that have given birth to the diversified features of that surface, we may, therefore, ignore the influence of air and water action and confine our reasoning to igneous phenomena alone: our task in this matter, it is hardly necessary to remark, is materially simplified thereby.

CHAPTER VI.

THE GENERAL ASPECT OF THE LUNAR SURFACE.

WE have now reached that stage of our subject at which it behoves us to repair to the telescope for the purpose of examining and familiarising ourselves with the various classes of detail that the lunar surface presents to our view.

That the moon is not a smooth sphere of matter is a fact that manifested itself to the earliest observers. The naked eye perceives on her face spots exhibiting marked differences of illumination. These variations of light and shade, long before the invention of the telescope, induced the belief that she possessed surface irregularities like those that diversify the face of the earth, and from analogy it was inferred that seas and continents alternated upon the lunar globe. It was evident, from the persistence and invariability of the dusky markings, that they were not due to atmospheric peculiarities, but were veritable variations in the character or disposition of the surface material. Fancy made pictures of these unchangeable spots : untutored gazers detected in them the indications of a human countenance, and perhaps the earliest map of the moon was a rough reproduction of a man's face, the eyes, nose and mouth representing the more salient spots discernible upon the lunar disc. Others recognised in these spots the configuration of a human form, head, arms and legs complete, which a French superstition that lingers to the present day held to be the image of Judas Iscariot transported to the moon in punishment for his treason. Again, an Indian notion connects the lunar spots with a representation of a roebuck or a hare, and hence the Sanskrit names for the moon, *mrigadhara*, a roebuck-bearer, and *'sa'sabhrit*, a hare-bearer. Of these similitudes the one which has the best pretensions to a rude accuracy is that first mentioned; for the resemblance of the full moon to a human countenance, wearing a painful or lugubrious expression, is very striking. Our illustration of the full moon (Plate III.) is derived from an actual photograph;[*] the relative intensities of light and shade are hence somewhat exaggerated; otherwise it represents the full moon very nearly as the naked eye sees it, and by gazing at the plate from a short distance,[†] the well-known features will manifest themselves, while they who choose may amuse themselves by arranging the markings in their imagination till they conform to the other appearances alluded to.

We may remark in passing that by one sect of ancient writers the moon was supposed to be a kind of mirror, receiving the image of the earth and reflecting it back to terrestrial spectators. Humboldt affirmed that this opinion had been preserved to his day as a popular belief among the people of Asia Minor. He says, " I was once very much astonished to hear a very well educated Persian from Ispahan, who certainly had never read a Greek book, mention when I showed him the moon's spots in a large telescope in Paris, this hypothesis as a widely diffused belief in his country : ' What we see in the moon,' said the Persian, ' is ourselves ; it is the map of our earth.' " Quite as extravagant an idea, though perhaps a more excusable one, was that held by some ancient philosophers, to the effect that the spots on the moon were the shadows of opaque bodies floating in space between it and the sun.

An observer watching the forms and positions of the lunar face-marks, from night to night and from lunation to lunation, cannot fail to notice the circumstance that they undergo no easily perceptible change of position with respect to the circular outline of the disc ; that in fact the face of the moon presented to our view is always the same, or very nearly so. If the moon had no orbital motion we should be led from the above phenomenon to conclude that she had no axial motion, no movement of

[*] For the original photograph from which this plate was produced, and for permission to reproduce it, we owe our acknowledgments to Warren De la Rue and Joseph Beck, Esquires.

[†] The proper distance for realising the conditions under which the moon itself is seen will be that at which our disc is just covered by a wafer about a quarter of an inch in diameter, held at arm's length. This will subtend an angle of about half a degree, which is nearly the angular diameter of the moon.

rotation; but when we consider the orbital motion in connection with the permanence of aspect, we are driven to the conclusion—one, however, which superficial observers have some difficulty in recognising—that the moon has an axial rotation equal in period to her orbital revolution. Since the moon makes the circuit of her orbit in twenty-seven days and one-third (more exactly 27d. 7h. 43m. 11s.) it follows that this is the time of her axial rotation, as referred to the stars, or as it would be made out by an observer located at a fixed position in space outside the lunar orbit. But if referred to the sun this period appears different; because the moon while revolving round the earth is, with the earth, circulating around the sun. Suppose the three bodies, moon, earth, and sun, to be in a line at a certain period of a lunation, as they are at full moon: by the time the moon has completed her twenty-seven days' journey around the earth, the latter will have moved along twenty-seven days' march of its orbit, which is about twenty-seven degrees of celestial longitude: the sun will apparently be that much distant from a straight line passing through earth and moon, and the moon must therefore move forward to overtake the sun before she can assume the full phase again. She will take something over two days to do this; hence the solar period of her revolution becomes more than twenty-nine days (to be exact, 29d. 12h. 44m. 2s. ·87). This is the length of a solar day upon the moon—the interval from one sunrise to another at any spot upon the equator of our satellite, and the interval between successive reappearances of the same phase to observers on the earth. The physical cause of the coincidence of times of rotation and revolution was touched upon in a previous chapter.

We have said that the moon continuously presents to us the same hemisphere. This is generally true, but not entirely so. Galileo, by long scrutiny, familiarised himself with every detail of the lunar disc that came within the limited grasp of his telescopes, and he recognised the fact that according as the position of the moon varied in the sky, so the aspect of her face altered to a slight degree; that certain regions at the edge of her disc alternately came in sight and receded from his view. He perceived, in fact, an *apparent* rocking to and fro of the globe of the moon; a sort of balancing or *libratory* motion. When the moon was near the horizon he could see spots upon her uppermost edge, which disappeared as she approached the zenith, or highest point of her nightly path; and as she neared this point, other spots, before invisible, came into view, near to what had been her lower edge. Galileo was not long in referring this phenomenon to its true cause. The centre of motion of the moon being the centre of the earth, it is clear that an observer on the surface of the latter, looks down upon the rising moon as from an eminence, and thus he is enabled to see more or less over or around her. As the moon increases in altitude, the line of sight gradually becomes parallel to the line joining the observer and the centre of the earth, and at length he looks her full in the face: he loses the full view and catches another side face view as she nears the horizon in setting. This phenomenon, occurring as it does, with a daily period, is known as the *diurnal libration*.

But a kindred phenomenon presents itself in another period, and from another cause. The moon rotates upon her axis at a speed that is rigorously uniform. But her orbital motion is not uniform, sometimes it is faster, and at other times slower than its average rate. Hence, the angle through which she moves along her orbit in a given time, now exceeds, and now falls short of the angle through which she turns upon her axis. Her visible hemisphere thus changes to an extent depending upon the difference between these orbital and axial angles, and the apparent balancing thus produced is called the *libration in longitude*. Then there is a *libration in latitude* due to the circumstance that the axis of the moon is not exactly perpendicular to the plane of her orbit; the effect of this inclination being, that we sometimes see a little more of the north than of the south polar regions of our satellite, and *vice versâ*.*

---

* The libratory movement has been taken advantage of, at the suggestion of Sir Chas. Wheatstone, for producing stereoscopic photographs of the moon. In the early days of stereoscopic photography the object to be photographed was placed upon a kind of turn-table, and, after a picture had been taken of it in one position, the table was turned through a small angle for the taking of the second picture; the two placed side by side then represented the object as it would have been seen by two eyes widely separated, or whose visual rays inclined at an angle equal to that through which the table was turned; and when the pictures were viewed through a stereoscope, they combined to produce the wonderful effect of solidity now familiar to every one. The moon, by its librations, imitates the turn-table movement; and, from a large number of photographs of her, taken at different points of her orbit and at different seasons of the year, it is possible to select two which, while they exhibit the same phase of illumination, at the same time present the requisite difference in the points of view from which they are taken

The extent of the moon's librations, taking them all and in combination into account, amounts to about seven degrees of arc of latitude or longitude upon the moon, both in the north-south and east-west directions. And taking into account the whole effect of them, we may conclude that our view of the moon's surface, instead of being confined to one half, is extended really to about four-sevenths of the whole area of the lunar globe. The remaining three-sevenths must for ever remain a *terra incognita* to the habitants of this earth, unless, indeed, from some catastrophe which it would be wild fancy to anticipate, a period of rotation should be given to the moon different from that which it at present possesses. Some highly fanciful theorists have speculated upon the possible condition of the invisible hemisphere, and have propounded the absurd notion that the opposite side of the moon is hollow, or that the moon is a mere shell; others again have urged that the hidden half is more or less covered with water, and others again that it is peopled with inhabitants. There is, however, no good reason for supposing that what we may call the back of the moon has a physical structure essentially different from the face presented towards us. So far as can be judged from the peeps that libration enables us to obtain, the same characteristic features (though of course with different details) prevail over the whole lunar surface.

The speculative ideas held by the philosophers of the pre-telescopic age, touching the causes which produced the inequalities of light and shade upon the moon, received their *coup de grâce* from the revelations of Galileo's glasses. Our satellite was one of the earliest objects, if not actually the first, upon which the Florentine turned his telescope; and he found that the inequalities upon her surface were due to differences

---

to give the effect of stereoscopicity when viewed binocularly. Mr. De la Rue, the father of celestial photography, has been enabled to produce several such pairs of pictures from the vast collection of lunar photographs that he has accumulated. Any one of these pairs of portraits, when stereoscopically combined, reproduces, to quote the words of Sir John Herschel, "*the spherical* form just as a giant might see it whose stature were such that the interval between his eyes should equal the distance between the place where the earth stood when one view was taken, and that to which it would have to be removed (our moon being fixed) to get the other. Nothing can surpass the impression of *real corporeal form* thus conveyed by some of these pictures as taken by Mr. De la Rue with his powerful reflector, the production of which (as a step in some sort taken by man outside of the planet he inhabits) is one of the most remarkable and unexpected triumphs of scientific art."

in its configuration analogous to the continents and islands, and (as might then have been thought) the seas of our globe. He could trace, even with his moderate means, the semblance of mountain-tops upon which the sun shone while their lower parts were in shadow, of hills that were brightly illuminated upon their sides towards the sun, of brightly shining elevations, and deeply shadowed depressions, of smooth plains, and regions of mountainous ruggedness. He saw that the boundary of sunlight upon the moon was not a clearly defined line, as it would be if the lunar globe were a smooth sphere, as the Aristotelians had asserted, but that the terminator was uneven and broken into an irregular outline. From these observations the Florentine astronomer concluded that the lunar world was covered not only with mountains like our globe, but with mountains whose heights far surpassed those existing upon the earth, and whose forms were strangely limited to circularity.

Galileo's best telescopes magnified only some thirty times, and the views which he thus obtained, must have been similar to those exhibited by the smaller photographs of the moon produced in late years by Mr. De la Rue and now familiar to the scientific public. Of course there is in the natural moon as viewed with a small telescope a vivid brilliancy which no art can imitate, and in photographs especially there is a tendency to exaggeration of the depths of shade in a lunar picture. This arises from the circumstance that various regions of the moon do not impress a chemically sensitized plate as they impress the retina of the eye. Some portions, notably the so-called "seas" of the moon, which to the eye appear but slightly duller than the brighter parts, give off so little *actinic* light that they appear as nearly black patches upon a photograph, and thus give an undue impression of the relative brightness of various parts of the lunar surface. Doubtless by sufficient exposure of the plate in the camera-telescope the dark patches might be rendered lighter, but in that case the more strongly illuminated portions, which after all are those most desirable to be preserved, would be lost by the effect which photographers understand as "solarization."

In speaking of a view of the moon with a magnifying power of thirty, it is necessary to bear in mind that the visible features will differ considerably with the diameter of the object-glass of the telescope to which this power is applied. The same details would not be seen alike with the same power upon an object-glass of 10 inches diameter and one of 2 inches. The superior illumination of the image in the former case would bring into view minute details that could not be perceived with the smaller aperture. He who would for curiosity wish to see the moon, or any other object, as Galileo saw it, must use a telescope of the same size and character in all respects as Galileo's: it will not do to put his magnifying power upon a larger telescope. With large telescopes, and low powers used upon bright objects like the moon, there is a blinding flood of light which tends to contract the pupil of the eye and prevent the passage of the whole of the pencil of rays coming through the eye-piece. Although this last result may be productive of no inconvenience, it is clearly a waste of light, and it points to a rule that the lowest power that a telescope should bear is that which gives a pencil of light equal in diameter to the pupil of the eye under the circumstances of brightness attendant upon the object viewed. In observing faint objects this point assumes more importance, since it is then necessary that all available light should enter the pupil. The thought suggests itself that an artificial enlargement of the pupil, as by a dose of belladonna, might be of assistance in searching for faint objects, such as nebulæ and comets: but we prefer to leave the experiment for those to try who pursue that branch of astronomical observation.

A merely cursory examination of the moon with the low power to which we have alluded is sufficient to show us the more salient features. In the first place we cannot help being struck with the immense preponderance of circular or craterform asperities, and with the general tendency to circular shape which is apparent in nearly all the lunar surface markings; for even the larger regions known as the "seas" and the smaller patches of the same character seem to repeat in their outlines the round form of the craters. It is at the boundary of sunlight on the lunar globe that we see these craterform spots to the best advantage, as it is there that the rising or setting sun casts long shadows over the lunar landscape, and brings elevations and asperities into bold relief. They vary greatly in size, some are so large as to bear an estimable proportion to the moon's diameter, and the smallest are so minute as to need the most powerful telescopes and the finest conditions of atmosphere to perceive them. It is doubtful whether the smallest of them have ever been seen, for there is no reason to doubt that there exist countless numbers that are beyond the revealing powers of our finest telescopes.

From the great number and persistent character of these circumvallations, Kepler was led to think that they were of artificial construction. He regarded them as pits excavated by the supposed habitants of the moon to shelter themselves from the long and intense action of the sun. Had he known their real dimensions, of which we shall have to speak when we come to describe them more in detail, he would have hesitated in propounding such a hypothesis; nevertheless it was, to a certain extent, justified by the regular and seemingly unnatural recurrence of one particular form of structure, the like of which is, too, so seldom met with as a structural feature of the surface of our own globe.

The next most striking features, revealed by a low telescopic power upon the moon, are the seemingly smooth plains that have the appearance of dusky spots, and that collectively cover a considerable portion—about two-thirds—of the entire disc. The larger of these spots retain the name of *seas*, the term having been given when they were supposed to be watery expanses, and having been retained, possibly to avoid the confusion inevitable from a change of name, after the existence of water upon the moon was disproved. Following the same order of nomenclature, the smaller spots have received the appellations of *lakes*, *bays*, and *fens*. We see that many of these "seas" are partially surrounded by ramparts or bulwarks which, under closer examination, and having regard to their real magnitude, resolve themselves into immense mountain chains. The general resemblance in form which the bulwarked plains thus exhibit to the circular craters of large size, would lead us to suppose that the two classes of objects had the same formative origin, but when we take into account the immense size of the former, and the process by which we infer the latter to have been developed, the supposition becomes untenable.

Another of the prominent features which we notice as highly curious, and in some phases of the moon—at about the time of full—the most remarkable of all, are certain bright lines that appear to radiate from some of the more conspicuous craters, and extend for hundreds of miles around. No selenological formations have so sorely puzzled observers as these peculiar streaks, and a great deal of fanciful theorizing has been bestowed upon them. As we are now only glancing at the moon, we do not enter upon explanations concerning them or any other class of details; all such will receive due consideration in their proper order in succeeding chapters.

We thus see that the classes of features observable upon the moon are not great in number: they may be summed up as *craters* and their central cones, *mountain chains*, with occasional isolated peaks, *smooth plains*, with more or less of irregularity of surface, and *bright radiating streaks*. But when we come to study with higher powers the individual examples of each class we meet with considerable diversity. This is especially the case with the craters, which appear under very numerous variations of the one order of structure, viz., the ring-form. A higher telescopic power shows us that not only do these craters exist of all magnitudes within a limit of largeness, but seemingly with no limit of smallness, but that in their structure and arrangement they present a great variety of points of difference. Some are seen to be considerably elevated above the surrounding surface, others are basins hollowed out of that surface and with low surrounding ramparts; some are merely like walled plains or amphitheatres with flat plateaux, while the majority have their lowest point of hollowness considerably below the general level of the surrounding surface; some are isolated upon the plains, others are aggregated into a thick crowd, and overlapping and intruding upon each other; some have elevated peaks or cones in their centres, and some are without these central cones, while the plateaux of others again contain several minute craters instead; some have their ramparts whole and perfect, others have them breached or malformed, and many have them divided into terraces, especially on their inner sides.

In the plains, what with a low power appeared smooth as a water surface becomes, under greater magnification, a rough and furrowed area, here gently undulated and there broken into ridges and declivities, with now and then deep rents or cracks extending for miles and spreading like river-beds into numerous ramifications. Craters of all sizes and classes are scattered over the plains ; these appear generally of a different tint to the surrounding surface, for the light reflected from the plains has been observed to be slightly tinged with colour. The tint is not the same in all cases : one large sea has a dingy greenish tinge, others are merely grey, and some others present a pale reddish hue. The cause of this diversity of colour is mysterious ; it has been supposed to indicate the existence of vegetation of some sort ; but this involves conditions that we know do not exist.

The mountains, under higher magnification, do not present such diversity of formation as the craters, or at least the points of difference are not so apparent ; but they exhibit a plentiful variety of combinations. There are a few perfectly isolated examples that cast long shadows over the plains on which they stand like those of a towering cathedral in the rising or setting sun. Sometimes they are collected into groups, but mostly they are connected into stupendous chains. In one of the grandest of these chains, it has been estimated that a good telescope will show 3000 mountains clustered together, without approach to symmetrical order. The scenery which they would present, could we get any other than the "bird's eye view" to which we are confined, must be imposing in the extreme, far exceeding in sublime grandeur anything that the Alps or the Himalayas offer ; for while on the one hand the lunar mountains equal those of the earth in altitude, the absence of an atmosphere, and consequently of the effects produced thereby, must give rise to alternations of dazzling light and black depths of shade combining to form panoramas of wild scenery that, for want of a parallel on earth, we may well call unearthly. But we are debarred the pleasure of actually contemplating such pictures by the circumstance that we look *down* upon the mountain tops and into the valleys, so that the great height and close aggregation of the peaks and hills are not so apparent. To compare the lunar and terrestrial mountain scenery would be "to compare the different views of a town seen from the car of a balloon with the more interesting prospects by a progress through the streets." Some of the peculiarities of the lunar scenery we have, however, endeavoured to realize in a subsequent Chapter.

A high power gives us little more evidence than a low one upon the nature of the long bright streaks that radiate from some of the more conspicuous craters, but it enables us to see that those streaks do not arise from any perceptible difference of level of the surface—that they have no very definite outline, and that they do not present any sloping sides to catch more sunlight, and thus shine brighter, than the general surface. Indeed, one great peculiarity of them is that they come out most forcibly where the sun is shining perpendicularly upon them; hence they are best seen where the moon is at full, and they are not visible at all at those regions upon which the sun is rising or setting. We also see that they are not diverted by elevations in their path, as they traverse in their course craters, mountains, and plains alike, giving a slight additional brightness to all objects over which they pass, but producing no other effect upon them. To employ a commonplace simile, they look as though, after the whole surface of the moon had assumed its final configuration, a vast brush charged with a whitish pigment had been drawn over the globe in straight lines radiating from a central point, leaving its trail upon everything it touched, but obscuring nothing.

Whatever may be the cause that produces this brightness of certain parts of the moon without reference to configuration of surface, this cause has not been confined to the formation of the radiating lines, for we meet with many isolated spots, streaks and patches of the same bright character. Upon some of the plains there are small areas and lines of luminous matter possessing peculiarities similar to those of the radiating streaks, as regards visibility with the high sun, and invisibility when the solar rays fall upon them horizontally. Some of the craters also are surrounded by a kind of aureole of this highly reflective matter. A notable specimen is that called *Linné*, concerning which a great hue and cry about change of appearance and inferred continuance of volcanic action on the moon was raised some years ago. This object is an insignificant little crater of about a mile or two in diameter, in the centre of an ill-defined spot of the character referred to, and about eight or ten miles in diameter. With a low sun the crater alone is visible by its shadow; but as the luminary rises the shadow shortens and becomes all but invisible, and then the white spot shines forth. These alternations, complicated by variations of atmospheric condition, and by the interpretations of different observers, gave rise to statements of somewhat exaggerated character to the effect that considerable changes, of the nature of volcanic eruptions, were in progress in that particular region of the moon.

In the foregoing remarks we have alluded somewhat indefinitely to high powers; and an enquiring but unastronomical reader may reasonably demand some information upon this point. It might have been instructive to have cited the various details that may be said to come into view with progressive increases of magnification. But this would be an all but impossible task, on account of the varying conditions under which all astronomical observations must necessarily be made. When we come to delicate tests, there are no standards of telescopic power and definition. Assuming the instrument to be of good size and high optical character, there is yet a powerful influant of astronomical definition in the atmosphere and its variable state. Upon two-thirds of the clear nights of a year the finest telescopes cannot be used to their full advantage, because the minute flutterings resulting from the passage of the rays of light through moving strata of air of different densities are magnified just as the image in the telescope is magnified, till all minute details are blurred and confused, and only the grosser features are left visible. And supposing the telescope and atmosphere in good state, there is still an important point, the state of the observer's eye, to be considered. After all it is the eye that sees, and the best telescopic assistance to an untrained eye is of small avail. The eye is as susceptible of education and development as any other organ; a skilful and acute observer is to a mere casual gazer what a watchmaker would be to a ploughman, a miniature painter to a whitewasher. This fact is not generally recognized; no man would think of taking in hand an engraver's burin, and expecting on the instant to use it like an adept, or of going to a smithy and without previous preparation trying to forge a horse-shoe. Yet do folks enter observatories with uneducated eyes, and expect at once to realise all the wonderful things that their minds have pictured to themselves from the perusal of astronomical books. We have over and over again remarked the dissatisfaction which attends the first looks of novices through a powerful telescope. They anticipate immediately beholding wonders, and they are disappointed at finding how little they can see, and how far short the sight falls of what they had expected. Courtesy at times leads them to express wonder and surprise, which it is easy to see is not really felt, but sometimes honesty compels them to give expression to their disappointment. This arises from the simple fact that their eyes are not fit for the work which is for the moment imposed upon them; they know not what to look for, or how to look for it. The first essay at telescopic gazing, like first essays generally, serves but to teach us our incapability.

To a tutored eye a great deal is visible with a comparatively low power, and practised observers strive to use magnifying powers as low as possible, so as to diminish, as far as may be, the evils arising from an untranquil atmosphere. With a power so small as 30 or 40, many exceedingly delicate details on the moon are visible to an eye that is familiar with them under higher powers. With 200 we may say that every ordinary detail will come out under favourable conditions; but when minute points of structure, mere nooks and corners as it were, are to be scrutinised, 300 may be used with advantage. Another hundred diameters almost passes the practical limit. Unless the air be not merely fine, but superfine, the details become "clothy" and tremulous; the extra points brought out by the increased power are then only caught by momentary glimpses, of which but a very few are obtained during a lengthy period of persistent scrutiny. We may set down 250 as the most useful, and 350 the utmost effective power that can be employed upon the particular work of which we are treating. Could every detail on the moon be thoroughly and reliably represented as this amount of magnification shows it, the result would leave little to be wished for.

But it may be asked by some, what is the absolute effect of such powers as those we have spoken of, in bringing the moon apparently nearer to our eyes? and what is the actual size of the smallest object visible under the most favourable circumstances? A linear mile upon the moon corresponds to an angular interval of 0·87 of a second; this refers to regions about the centre of the disc; near the circumference the foreshortening makes a difference, very great as the edge is approached. Perhaps the smallest angle that the eye can without assistance appreciate is half a minute; that is to say, an object that subtends to the eye an arc of less than half a minute can scarcely be seen.* Since there are 60 seconds in a minute, it follows that we must magnify a spot a second in diameter upon the moon thirty times before we can see it; and since a second represents rather more than a mile, really about 2000 yards, on the moon, as seen from the earth, the smallest object visible with a power of 30 will be this number of yards in diameter or breadth. To see an object 200 yards across, we should require to magnify it 300 times, and this would only bring it into view as a point; 20 yards would require a power of 3000, and 1 yard 60,000 to effect the same thing. Since, as we have said, the highest practicable power with our present telescopes, and at ordinary terrestrial elevations, is 350, or for an extreme say 400, it is evident that the minutest lunar object or detail of which we can perceive as a point must measure about 150 yards : to see the form of an object, so as to discriminate whether it be round or square, it would require to be probably twice this size ; for it may be safely assumed that we cannot perceive the outline of an object whose average breadth subtends a less angle than a minute.

Arago put this question into another shape :—The moon is distant from us 237,000 miles (mean). A magnifying power of a thousand would show us the moon as if she were distant 237 miles from the naked eye.

                2000 would bring her within 118 miles.
                4000   „   „   „   59   „
                6000   „   „   „   39   „

Mont Blanc is visible to the naked eye from Lyons, at the distance of about 100 miles; so that to see the mountains of the moon as Mont Blanc is seen from Lyons would require the impracticable power of 2500.

---

* This is a point of some uncertainty. Dr. Young stated (Lectures Vol. II. p. 575) that "a minute is perhaps nearly the smallest interval at which two objects can be distinguished, although a line subtending only a tenth of a minute in breadth may sometimes be perceived as a single object."

CHAPTER VII.

TOPOGRAPHY OF THE MOON.

It is scarcely necessary to seek the reasons which prompted astronomers, soon after the invention of the telescope, to map the surface features of the moon. They may have considered it desirable to record the positions of the spots upon her disc, for the purpose of facilitating observations of the passage of the earth's shadow over them in lunar eclipses; or they may have been actuated by a desire to register appearances then existing, in order that if changes took place in after years these might be readily detected. Scheiner was one of the earliest of lunar cartographers; he worked about the middle of the seventeenth century; but his delineations were very rough and exaggerated. Better maps—the best of the time, according to an old authority—were engraved by one Mellan, about the years 1634 or 1635. At about the same epoch, Langreen and Hevelius were working upon the same subject. Langreen executed some thirty maps of portions of the moon, and introduced the practice of naming the spots after philosophers and eminent men. Hevelius spent several years upon his task, the results of which he published in a bulky volume containing some 50 maps of the moon in various phases, and accompanied by 500 pages of letter-press. He rejected Langreen's system of nomenclature, and called the spots after the seas and continents of the earth to which he conceived they bore resemblance. Riccioli, another selenographer, whose map was compiled from observations made by Grimaldi, restored Langreen's nomenclature, but he confined himself to the names of eminent astronomers, and his system has gained the adhesion of the map-makers of later times. Cassini prepared a large map from his own observations, and it was

K

engraved about the year 1692. It appears to have been regarded as a standard work, for a reduced copy of it was repeatedly issued with the yearly volumes of the *Connaissance desTemps*, (the "Nautical Almanack" of France) some time after its publication. These small copies have no great merit: the large copper plate of the original was, we are told by Arago, who received the statement from Bouvard, sold to a brazier by a director of the French Government Printing-Office, who thought proper to disembarrass the stores of that establishment, by ridding them of what he considered lumber! La Hire, Mayer, and Lambert, followed during the succeeding century, in this branch of astronomical delineation. At the commencement of the present century, the subject was very earnestly taken up by the indefatigable Schroeter, who, although he does not appear to have produced a complete map, produced a topograph of the moon in ·a large series of partial maps and drawings of special features. Schroeter was a fine observer, but his delineations show him to have been an indifferent draughtsman. Some of his drawings are but the rudest representations of the objects he intended to depict; many of the bolder features of conspicuous objects are scarcely recognizable in them. A bad artist is as likely to mislead posterity as a bad historian, and it cannot be surprising if observers of this or future generations, accepting Schroeter's drawings as faithful representations, should infer from them remarkable changes in the lunar details. It is much to be regretted that Schroeter's work should be thus depreciated. Lohrman of Dresden, was the next cartographer of the moon; in 1824 he put forth a small but very excellent map of 15 inches diameter, and published a book of descriptive text, accompanied by sectional charts of particular areas. His work, however, was eclipsed by the great one which we owe to the joint energy of MM. Beer and Maedler, and which represents a stupendous amount of observing work carried on during several years prior to 1836, the date of their publication. The long and patient labour bestowed upon their map and upon the measures on which it depends, deserve the highest praise which those conversant with the subject can bestow, and it must be very long before their efforts can be superseded.

Beer and Maedler's map has a diameter of 37 inches : it represents the phase of the moon visible in the condition of mean libration. The details were charted by a careful process of triangulation. The disc was first divided into "triangles of the first order," the points of which (conspicuous craters) were accurately laid down by reference to the edges of the disc: one hundred and seventy-six of these triangles, plotted accurately upon an orthographic projection of the hemisphere, formed the reliable basis for their charting work. From these a great number of "points of the second order" were laid down, by measuring their distance and angle of position with regard to points first established. The skeleton map thus obtained was filled up by drawings made at the telescope: the diameters of the measureable craters being determined by the micrometer.

Beer and Maedler also measured the heights of one thousand and ninety-five lunar mountains and crater-summits: the resulting measures are given in a table contained in the comprehensive text-book which accompanies their map. These heights are found by one of two methods, either by measuring the length of the shadow which the object casts under a known elevation of the sun above its horizon, or by measuring the distance between the illuminated point of the mountain and the "terminator" in the following manner. In the annexed figure (Fig. 15) let the circle represent the moon and M a mountain upon it: let S A be the line of direction of the sun's rays, passing the normal surface of the moon at A and just tipping the mountain top. A will be the terminator, and there will be darkness between it and the star-like mountain summit M. The distance between A and M is measured: the distance A B is known, for it is the moon's radius. And since the line S M is a tangent to the circle the angle B A M is a right angle. We know the length of its two sides AB, AM, and we can therefore by the known properties of the right-angled triangle find the length of the hypothenuse BM: and since BM is made up of the radius BA plus the mountain height, we have only to subtract the moon's radius from the ascertained whole length of the hypothenuse and we have the height of the mountain. MM. Beer and Maedler exhibited their measures in French toises: in the heights we shall have occasion to quote, these have been turned into English feet, upon the assumption that the toise is equal to

6·39 English feet. The nomenclature of lunar features adopted by Beer and Maedler is that introduced by Riccioli : mountains and features hitherto undistinguished were named by them after ancient and modern

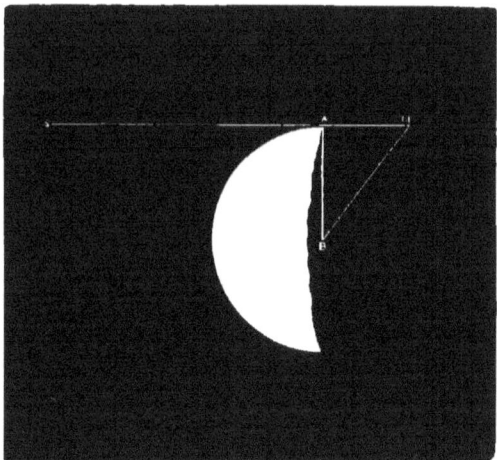

Fig. 15.

philosophers, in continuance of Riccioli's system, and occasionally after terrestrial features. Some minute objects in the neighbourhood of large and named ones were included under the name of the large one and distinguished by Greek or Roman letters.

The excellent map resulting from the arduous labours of these astronomers is simply a map : it does not pretend to be a picture. The asperities and depressions are symbolized by a conventional system of shading and no attempt is made to exhibit objects as they actually appear in the telescope. A casual observer comparing details on the map with the same details on the moon itself would fail to identify or recognize them except where the features are very conspicuous. Such an observer would be struck by the shadows by which the lunar objects reveal themselves : he would get to know them mostly by their shadows,

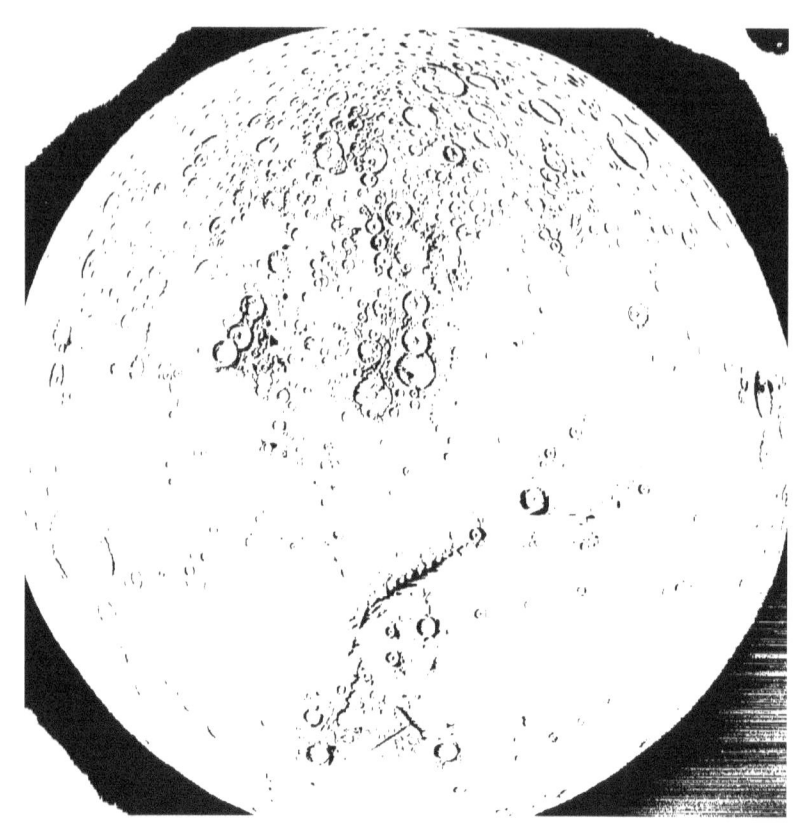

SKELETON MAP OF MOON
TO ACCOMPANY PICTURE MAP, CHAP. VII

PLATE

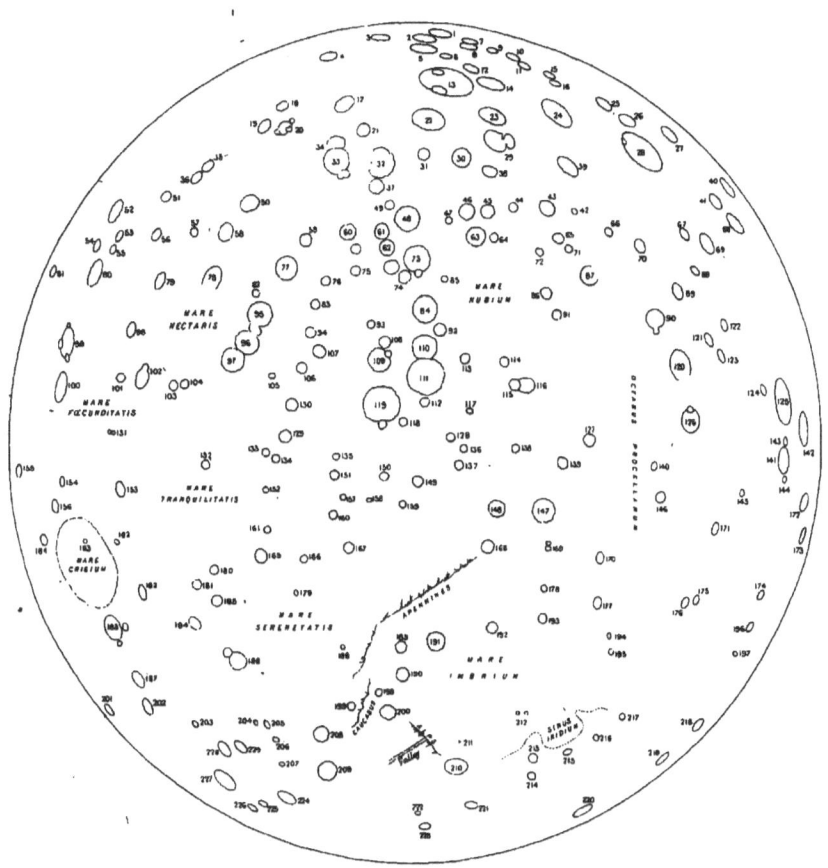

TOPOGRAPHY OF THE MOON.

since it is mainly by those that their forms are revealed to a terrestrial observer. But such a map as that under notice indicates no shadows, and objects have to be identified upon it rather by their positions with regard to one another or to the borders of the moon than by any notable features they actually present to view. This inconvenience occurred to us in our early use of Beer and Maedler's chart, and we were induced to prepare for ourselves a map in which every object is shown somewhat, if imperfectly, as it actually appears at some period of a lunation. This was done by copying Beer and Maedler's outlines and filling them up by appropriate shading. To do justice to our task we enlarged our map to a diameter of six feet. Upon a circle of this diameter the positions and dimensions of all objects were laid down from the German original. Then from our own observations we depicted the general aspect of each object : and we so adjusted the shading that all objects should be shown under about the same angle of illumination—a condition which is never fulfilled upon the moon itself, but which we consider ourselves justified in exhibiting for the purpose of conveying a fair impression of how the various lunar objects actually appear at some one or other part of a lunation.

The picture-map thus produced has been photographed to the size convenient for this work : and in order to make it available for the identification of such objects as we may have occasion to refer to, we have placed around it a co-ordinate scale of arbitrary divisions by which any object can be found as by the latitude and longitude divisions upon a common geographical map. We have also prepared a skeleton map which includes the more conspicuous objects, and which faces the picture map. (Plates IV. and V.) The numbers on the skeleton map are those given in the second column of the accompanying table. The table also gives the co-ordinate positions of the various craters, the names of which are, for convenience of reference, printed in alphabetical order.

Name.	Number.	Map Ordinates.	Name.	Number.	Map Ordinates.
Abulfeda	107	30·0 120·7	Almanon	94	29·0 122·3
Agrippa	151	31·2 110·0	Alpetragius	92	40·8 122·4
Airy	93	34·7 123·0	Alphonsus	110	39·6 120·9
Albategnius	109	35·5 119·7	Apianus	62	33·6 129·3
Aliacensis	61	35·8 131·0	Apollonius	154	6·5 109·5

70   THE MOON.   [CHAP. VII.

Name.	Number.	Map Ordinates.	Name.	Number.	Map Ordinates.
Arago	152	24·7 108·7	Diophantus	194	55·5 96·3
Archimedes	191	40·3 95·8	Doppelmayer	70	58·6 129·6
Aristarchus	176	62·3 99·2			
Aristillus	190	37·0 93·3	Eneke	140	59·7 110·6
Aristotle	209	30·0 84·6	Endymion	227	20·6 83·8
Arzachael	84	39·5 124·0	Epigenes	223	39·0 79·5
Atlas	228	20·7 86·6	Erastothenes	168	44·6 104·0
Autolycus	180	36·8 95·5	Eudoxus	208	29·7 88·0
Azophi	76	30·7 126·8			
			Fabricius	35	20·0 130·8
Bacon	17	32·5 142·0	Fernelius	37	35·1 134·8
Baily	207	26·0 85·4	Firmicus	156	5·8 107·7
Barocius	34	31·8 139·5	Flamsteed	126	62·8 114·5
Bessel	179	27·4 100·1	Fontana	122	65·9 123·0
Bettinus	11	48·8 144·9	Fontenelle	221	43·0 81·3
Bianchini	215	51·6 86·3	Fourier	67	62·5 130·7
Billy	121	64·3 121·4	Fracastorius	78	20·5 127·0
Blancanus	12	43·7 144·8	Furnerius	52	11·7 133·0
Bonpland	116	48·5 117·6			
Borda	56	15·2 131·0	Gambart	138	47·2 112·2
Boscovich	160	31·1 106·8	Gartner	224	26·5 82·3
Bouvard	40	66·6 134·3	Gassendi	90	59·7 123·3
Briggs	196	68·0 97·2	Gauricus	46	43·5 132·5
Bullialdus	86	50·1 125·5	Gauss	201	10·3 90·3
Burg	206	25·5 87·5	Gay Lussac	169	50·1 103·8
			Geber	83	29·6 124·8
Calippus	199	32·4 90·3	Geminus	187	13·0 93·0
Campanus	71	52·3 120·0	Gérard	218	63·7 88·8
Capella	104	17·8 118·0	Goclenius	101	11·8 118·5
Capuanus	43	50·5 132·8	Godin	135	31·3 111·7
Casatus	7	43·7 147·0	Grimaldi	125	70·8 116·3
Cassini	200	35·5 89·7	Grueınberger	6	41·4 145·8
Catherina	95	24·7 124·0	Guériké	114	46·5 119·6
Cavalerius	144	71·2 109·5	Guttemberg	102	13·9 118·3
Cavendish	88	63·5 127·4			
Cichus	44	47·3 132·8	Hainzel	39	52·3 136·7
Clavius	13	41·8 143·5	Hansteen	123	63·5 119·9
Cleomides	183	10·7 97·0	Hase	54	9·8 129·5
Colombo	98	12·8 122·7	Heinsius	38	45·5 136·0
Condamine	214	48·7 84·2	Helicon	212	48·0 89·6
Condorcet	164	4·5 104·7	Hell	47	41·7 131·6
Copernicus	147	49·8 107·0	Hercules	229	22·3 86·7
Cyrillus	96	23·5 121·3	Herodotus	175	63·2 99·6
			Herschel	112	39·3 116·2
Damoiseau	124	60·2 117·2	Hesiodus	64	45·8 130·0
Davy	113	43·2 119·8	Hevelius	141	71·5 111·3
Deambrel	120	26·8 113·5	Hippalus	87	54·0 127·0
Delisle	195	55·7 95·2	Hommel	20	26·8 140·0
Descartes	106	28·5 119·3	Hyginus	158	33·6 108·0

TOPOGRAPHY OF THE MOON.

Name.	Number.	Map Ordinates.		Name.	Number.	Map Ordinates.	
Inghirami	27	61·3	138·0	Olbers	172	73·0	107·7
Isidorus	103	16·7	118·0				
				Pallas	149	38·6	109·5
Kant	105	25·8	118·5	Parrot	108	35·8	121·6
Kepler	146	60·0	106·0	Petavius	80	9·5	127·5
Kies	72	49·7	128·8	Phocylides	25	55·5	141·6
Kircher	10	47·5	145·8	Piazzi	41	65·0	133·5
Klaproth	8	43·5	146·7	Picard	163	8·3	104·7
				Piccolomini	58	21·7	131·0
La Caille	74	37·5	126·8	Pico	211	41·9	87·3
Lagrange	68	67·0	131·3	Pitatus	63	44·1	130·2
La Hire	177	54·3	99·3	Plana	205	24·8	88·8
Lalande	117	43·4	115·3	Plato	210	41·8	84·8
Lambert	193	49·6	97·8	Playfair	75	33·5	127·5
Landsberg	127	54·0	113·0	Pliny	165	24·2	103·4
Langreen	100	6·3	117·7	Poisson	60	32·8	131·0
Letronne	120	62·0	119·0	Polybius	82	24·5	125·6
Licetus	21	34·1	139·6	Pontanus	59	29·0	130·2
Lichtenberg	197	66·5	94·0	Posidonius	186	22·2	94·3
Linnæus	188	31·7	95·7	Proclus	162	11·4	104·5
Littrow	185	20·5	99·4	Ptolemy	111	39·5	118·2
Lohrman	143	71·3	112·6	Purbach	73	38·7	128·4
Longomontanus	23	45·7	140·6	Pythagoras	220	53·0	81·2
Lubiniczky	91	51·3	123·5	Pytheas	178	49·7	100·4
Macrobius	182	13·7	100·2	Ramsden	42	52·9	132·5
Maginus	22	40·0	140·4	Reamur	118	37·3	114·6
Mairan	217	56·7	89·5	Reiner	145	67·3	108·5
Manilius	167	32·2	103·9	Reinhold	139	51·5	111·2
Manzinus	4	31·3	146·0	Repsold	219	60·2	85·7
Maraldi	181	18·6	100·8	Rheita	51	16·1	134·2
Marius	171	65·0	105·5	Riccioli	142	72·7	113·8
Maskelyne	132	19·5	111·0	Riccius	50	23·7	133·5
Mason	204	23·7	88·8	Ritter	134	26·0	111·6
Maupertius	213	48·7	85·8	Roemer	184	18·3	97·6
Maurolycus	33	31·8	137·0	Ross	161	25·0	105·3
Menelaus	165	28·3	103·0				
Mercator	65	51·4	130·2	Sabine	133	25·0	112·0
Mersenius	89	61·7	125·7	Sacrobosco	77	27·5	127·7
Messala	202	14·0	90·5	Santbech	79	15·7	126·8
Messier	131	10·8	114·0	Saussure	31	39·6	137·7
Metius	36	18·8	135·9	Scheiner	14	45·5	143·5
Moretus	5	39·5	146·5	Schickard	28	59·0	137·5
Moesting	128	41·6	113·2	Schiller	24	51·3	141·0
				Schroeter	137	42·3	110·7
Neander	57	18·7	131·0	Schubert	155	2·3	110·8
Nearchus	18	26·8	142·0	Segner	16	51·3	143·5
Newton	1	41·0	147·7	Seleucus	174	69·0	99·8
Nonius	49	36·5	133·2	Sharp	216	54·2	87·7

Name.	Number.	Map Ordinates.	Name.	Number.	Map Ordinates.
Short	2	39·7 147·4	Tycho	30	43·0 142·3
Silberschlag	157	32·0 108·1			
Simpelius	3	35·8 147·7	Ukert	159	37·1 107·5
Snell	55	11·3 129·6			
Soemmering	136	42·8 112·2	Vasco de Gama	173	72·8 104·9
Stadius	148	45·6 107·0	Vendelinus	99	6·8 121·6
Stevinus	53	11·9 130·7	Vieta	69	64·3 129·7
Stoefler	32	35·6 136·8	Vitello	66	55·8 130·7
Strabo	226	23·2 81·6	Vitruvius	180	20·1 102·0
Struve	203	18·3 88·7	Vlacq	19	25·0 140·1
Taruntius	153	11·7 109·0	Walter	48	37·8 131·0
Taylor	130	27·6 116·2	Wargentin	26	57·5 140·2
Thales	225	24·3 81·8	Werner	62	36·4 129·3
Thebit	85	40·8 126·8	Wilhelm Humboldt	81	5·7 127·8
Theophilus	97	22·3 120·0	Wilhelm I.	29	45·9 138·6
Timæus	222	38·3 80·8	Wilson	9	45·7 146·4
Timocharis	192	45·1 97·0	Wurzelbauer	45	45·0 132·6
Tobias Mayer	170	54·5 103·0			
Triesnecker	150	35·5 109·8	Zuchius	15	50·7 144·2

The strong family likeness pervading the craters of the moon renders it unnecessary that we should attempt a description of each one of them or even of one in twenty. We have, however, thought that a few remarks upon the salient features of a few of the most important may be acceptable in explanation of our illustrative plates; and what we have to say of the few may be taken as representative of the many.

COPERNICUS, 147—(49·8—107·0). PLATE VIII.

This may deservedly be considered as one of the grandest and most instructive of lunar craters. Although its vast diameter (46 miles) is exceeded by others, yet, taken as a whole, it forms one of the most impressive and interesting objects of its class. Its situation, near the centre of the lunar disc, renders all its wonderful details, as well as those of its immediately surrounding objects, so conspicuous as to establish it as a very favourite object. Its vast rampart rises to upwards of 12,000 feet above the level of the plateau, nearly in the centre of which stands a magnificent group of cones, three of them attaining the height of upwards of 2400 feet.

The rampart is divided by concentric segmental terraced ridges, which present every appearance of being enormous landslips, resulting from the crushing of their over-loaded summits, which have slid down in vast segments and scattered their débris on to the plateau. Corresponding vacancies in the rampart may be observed from whence these prodigious masses have broken away. The same may be noticed, although in a somewhat modified degree, around the exterior of the rampart. In order to approach a realization of the sublimity and grandeur of this magnificent example of a lunar volcanic crater, our reader would do well to endeavour to fix his attention on its enormous magnitude and attempt to establish in his mind's eye a correct conception of the scale of its details as well as its general dimensions, which, as they so prodigiously transcend those of the largest terrestrial volcanic craters, require that our ideas as to magnitude of such objects should be, so to speak, educated upon a special standard. It is for this reason we are anxious our reader, when examining our illustrations, should constantly refer the objects represented in them to the scale of miles appended to each plate, otherwise a just and true conception of the grandeur of the objects will escape him.

Copernicus is specially interesting, as being evidently the result of a vast discharge of molten matter which has been ejected at the focus or centre of disruption of an extensively upheaved portion of the lunar crust. A careful examination of the crater and the district around it, even to the distance of more than 100 miles on every side, will supply unmistakable evidence of the vast extent and force of the original disruption, manifested by a wonderfully complex reticulation of bright streaks which diverge in every direction from the crater as their common centre. These streaks do not appear on our plate, nor are they seen upon the moon except at and near the full phase. They show conspicuously, however, by their united lustre on the full moon, Plate III. Every one of those bright streaks, we conceive, is a record of what was originally a crack or chasm in the solid crust of the moon, resulting from some vastly powerful upheaving agency over the site of whose focus of energy Copernicus stands. The cracking of the crust must have been followed by the ejection of subjacent molten matter up through the reticulated cracks; this, spreading somewhat on either side of them, has left these bright streaks as a visible record of the force and extent of the upheaval;

while at the focus of disruption from whence the cracks diverge, the grand outburst appears to have taken place, leaving Copernicus as its record and result.

Many somewhat radial ridges or spurs may be observed leading away from the exterior banks of the great rampart. These appear to be due to the more free egress which the extruded matter would find near the focus of disruption. The spur-ridges may be traced fining away for fully 100 miles on all sides, until they become such delicate objects as to approach invisibility. Several vast open chasms or cracks may be observed around the exterior of the rampart. They appear to be due to some action subsequent to the formation of the great crater—probably the result of contraction on the cooling of the crust, or of a deep-seated upheaval long subsequent to that which resulted in the formation of Copernicus itself, as they intersect objects of evidently prior formation.

Under circumstances specially favourable for "fine vision," for upwards of 70 miles on all sides around Copernicus, myriads of comparatively minute but perfectly-formed craters may be observed. The district on the south-east side is specially rich in these wonderfully thickly scattered craters, which we have reason to suppose stand over or upon the reticulated bright streaks; but, as the circumstances of illumination which are requisite to enable us to detect the minute craters are widely adverse to those which render the bright streaks visible, namely, nearly full moon for the one and gibbous for the other, it is next to impossible to establish the fact of coincidence of the sites of the two by actual simultaneous observation.

At the east side of the rampart, multitudes of these comparatively minute craters may also be detected, although not so closely crowded together as those on the west side; but among those on the east may be seen myriads of minute prominences roughening the surface; on close scrutiny these are seen to be small mounds of extruded matter which, not having been ejected with sufficient energy to cause the erupted material to assume the crater form around the vent of ejection, have simply assumed the mound form so well known to be the result of volcanic ejection of moderate force.

Were we to select a comparatively limited portion of the lunar surface abounding in the most unmistakable evidence of volcanic action in every variety that can characterize its several phases, we could not choose one yielding in all respects such instructive examples as Copernicus and its immediate surroundings.

GASSENDI, 90—(59·7—123·3). Frontispiece.

An interesting crater about 54 miles diameter; the height of the most elevated portion of the surrounding wall from the plateau being about 9600 feet. The centre is occupied by a group of conical mountains, three of which are most conspicuous objects and rise to nearly 7000 feet above the level of the plateau. As in other similar cases, these central mountains are doubtless the result of the expiring effort of the eruption which had formed the great circular wall of the crater. The plateau is traversed by several deep cracks or chasms nearly one mile wide.

Both the interior and exterior of the wall of the crater are terraced with the usual segmental ridges or landslips. A remarkable detached portion of the interior bank is to be seen on the east side, while on the west exterior of the wall may be seen an equally remarkable example of an outburst of lava subsequent to the formation of the wall or bank of the crater; it is of conical form and cannot fail to secure the attention of a careful observer.

Interpolated on the north wall of the crater may be seen a crater of about 18 miles diameter which has burst its bank in towards the great crater, upon whose plateau the lava appears to have discharged itself.

The neighbourhood of Gassendi is diversified by a vast number of mounds and long ridges of exudated matter, and also traversed by enormous chasms and cracks, several of which exceed one mile wide and are fully 100 miles in length, and, as is usual with such cracks, traverse plain and mountain alike, disregarding all surface inequalities.

Numbers of small craters are scattered around; the whole forming an interesting and instructive portion of the lunar surface.

EUDOXUS, 208 (29·7—88·0), AND ARISTOTLE, 209 (30·0—84·6). PLATE X.

Two gigantic craters, Eudoxus being nearly 35 miles in diameter and upwards of 11,000 feet deep, while Aristotle is about 48 miles in diameter, and about 10,000 feet deep (measuring from the summit of the rampart to the plateau). These two magnificent craters present all the true volcanic characteristics in a remarkable degree. The outsides, as well as the insides of their vast surrounding walls or banks display on the grandest scale the landslip feature, the result of the over-piling of the ejected material, and the consequent crushing down and crumbling of the substructure. The true eruptive character of the action which formed the craters is well evinced by the existence of the groups of conical mountains which occupy the centres of their circular plateaux, since these conical mountains, there can be little doubt, stand over what were once the vents from whence the ejected matter of the craters was discharged.

On the west side of these grand craters may be seen myriads of comparatively minute ones (we use the expression "comparatively minute," although most of them are fully a mile in diameter). So thickly are these small craters crowded together, that counting them is totally out of the question ; in our original notes we have termed them "Froth craters" as the most characteristic description of their aspect.

The exterior banks of Aristotle are characterized by radial ridges or spurs : these are most probably the result of the flowing down of great currents of very fluid lava. To the east of the craters some very lofty mountains of exudation may be seen, and immediately beyond them an extensive district of smaller mountains of the same class, so thickly crowded together as under favourable illumination to present a multitude of brilliant points of light contrasted by intervening deep shade. On the west bank of Aristotle a very perfect crater may be seen, 27 miles in diameter, having all the usual characteristic features.

About 40 miles to the east of Eudoxus there is a fine example of a crack or fissure extending fully 50 miles—30 miles through a plain, and the remaining 20 miles cutting through a group of very lofty mountains. This great crack is worthy of attention, as giving evidence

TOPOGRAPHY OF THE MOON.

of the deep-seated nature of the force which occasioned it inasmuch as it disregards all surface impediments, traversing plain and group of mountains alike.

There are several other features in and around these two magnificent craters well worthy of careful observation and scrutiny, all of them excellent types of their respective classes.

TRIESNEKER, 150 (35·5—109·8). PLATE XI.

A fine example of a normal lunar volcanic crater, having all the usual characteristic features in great perfection. Its diameter is about 20 miles, and it possesses a good example of the central cone and also of interior terracing.

The most notable feature, however, in connection with this crater, and on account of which we have chosen it as a subject for one of our illustrations, is the very remarkable display of chasms or cracks which may be seen to the west side of it. Several of these great cracks obviously diverge from a small crater near the west external bank of the great one, and they subdivide or branch out, as they extend from the apparent point of divergence, while they are crossed or intersected by others. These cracks or chasms (for their width merits the latter appellation) are nearly one mile broad at the widest part, and after extending to fully 100 miles, taper away till they become invisible. Although they are not test objects of the highest order of difficulty, yet to see them with perfect distinctness requires an instrument of some perfection and all the conditions of good vision. When such are present, a keen and practised eye will find many details to rivet its attention, among which are certain portions of the edges of these cracks or chasms which have fallen in and caused interruptions to their continuity.

THEOPHILUS, 97 (22·3—120·0). CYRILLUS, 96 (23·5—121·3). CATHARINA, 95 (24·7—124·0). PLATE XII.

These three magnificent craters form a very conspicuous group near the middle of the south-east quarter of the lunar disc.

Their respective diameters and depths are as follows :—

Theophilus, 64 miles diameter; depth of plateau from summit of crater wall, 16,000 feet; central cone, 5200 feet high.

Cyrillus, 60 miles diameter; depth of plateau from summit of crater wall, 15,000 feet; central cone, 5800 feet high.

Catharina, 65 miles diameter; depth of plateau from summit of crater wall, 13,000 feet; centre of plateau occupied by a confused group of minor craters and débris.

Each of these three grand craters is full of interesting details, presenting in every variety the characteristic features which so fascinate the attention of the careful observer of the moon's wonderful surface, and affording unmistakable evidence of the tremendous energy of the volcanic forces which at some inconceivably remote period piled up such gigantic formations.

Theophilus by its intrusion within the area of Cyrillus shows in a very striking manner that it is of comparatively more recent formation than the latter crater. There are many such examples in other parts of the lunar disc, but few of so very distinct and marked a character.

The flanks or exterior banks of Theophilus, especially those on the west side, are studded with apparently minute craters, all of which when carefully scrutinized are found to be of the true volcanic type of structure; and minute as they are, by comparison, they would to a beholder close to them appear as very imposing objects; but so gigantic are the more notable craters in the neighbourhood, that we are apt to overlook what are in themselves really large objects. It is only by duly training the mind, as we have previously urged, so as ever to keep before us the vast scale on which the volcanic formations of the lunar surface are displayed, that we can do them the justice which their intrinsic grandeur demands. We trust that our illustrations may in some measure tend to educate the mind's eye, so as to derive to the full the tranquil enjoyment which results from the study of the manifestation of one of the Creator's most potent agencies in dealing with the materials of his worlds, namely, volcanic force. So rich in wonderful features and characteristic details is this magnificent group and its neighbourhood, that a volume might be filled in the attempt to do justice, by description, to objects so full of suggestive subject for study.

TOPOGRAPHY OF THE MOON.

PTOLEMY, ALPHONS, ARZACHAEL, ETC.—Plate XIII.
111 110 84

The portion of the lunar surface comprised within the limits of this illustration being situated nearly in the centre of the moon's disc, is very favourably placed for revealing the multitude of interesting features and details therein represented. They consist of every variety of volcanic crater from "Ptolemy," whose vast rampart is eighty-six miles diameter, down to those whose dimensions are, comparatively, so minute as to render them at the extreme limits of visibility.

Alphons and Arzachael, two of the next largest craters in our illustration, situated immediately above Ptolemy, are sixty and fifty-five miles diameter respectively, and are possessed, in a remarkable degree, of all the distinctive characteristic features of lunar craters, having magnificent central cones, lofty ragged ramparts, together with very striking manifestations of landslip formations as appear in the great segmental terraces within their ramparts, together with several minor craters interpolated on their plateau. "Alphons," the middle crater of this fine group, has its plateau specially distinguished by several cracks or chasms fully one mile wide, the direction or "strike" of which coincide in a very remarkable manner with several other similar cracks which form conspicuous features among the multitude of interesting details comprised within the limits of our illustration,—the most notable of these is an enormous straight cliff traversing the diameter of a low-ridged circular formation, seen in the upper right-hand corner of our plate. This great cliff is sixty miles long and from 1000 to 2000 feet high; it is a well-known object to lunar observers, and has been termed "The Railway," on account of its straightness as revealed by the distinct shadow projected by it on the plateau when seen under its sunrise aspect. The face of this vast cliff, although generally straight, is seen, when minutely scrutinized, to be somewhat serrated in its outline, while on its upper edge may be detected some very minute but perfectly formed craters. The existence of this remarkable cliff appears to be due either to an upheaval or a down-sinking of portion of the surface of the circular area across whose diameter it extends.

To the right-hand side of the cliff are two small craters, from the side of which a fine example of a crack may be detected passing through in its course a low dome-formed hill; this crack is parallel to the cliff, having in that respect the same general strike or parallel direction which so remarkably distinguishes the other cracks observable in this portion of the moon's surface.

On the left hand of this great cliff is situated a coneless crater, named "Thebit," on the right-hand rampart of which may be observed two small craters, the lesser of which is 2·75 miles diameter and has a central cone. We specially remark this fact, as it is the smallest lunar crater but one, in which we have, with perfect distinctness, detected a central cone. Not but that many smaller lunar craters exist possessed of this unmistakable evidence of their volcanic origin; but so minute are the specks of light which the central cones of such small craters reflect, that they, for that reason, most probably fail to reveal themselves.

PLATO, 210 (41·8—81·8). PLATE XIV.

This crater, besides being a conspicuous object on account of its great diameter, has many interesting details in and around it requiring a fine instrument and favourable circumstances to render them distinctly visible. The diameter of the crater is 70 miles; the surrounding wall or rampart varies in height from 4000 to upwards of 8000 feet, and is serrated with noble peaks which cast their black shadows across the plateau in a most picturesque manner, like the towers and spires of a great cathedral. Reference to our illustration will convey a very fair idea of this interesting appearance. On the north-east inside of the circular wall or rampart may be observed a fine example of landslip, or sliding down of a considerable mass of the interior side of the crater's wall. The landslip nature of this remarkable detail is clearly established by the fact of the bottom edge of the downslipped mass projecting in towards the centre of the plateau to a considerable extent. Other smaller landslip features may be seen, but none on so grand and striking a scale as the one referred to. A number of exceedingly minute craters may be detected on the surface of the plateau. The plateau itself is remarkable for its low reflective power, which causes it to look like a dingy spot when Plato is viewed with a small magnifying power. The exterior of the crater wall is remarkable for the rugged character of its formation, and forms a great contrast in that respect to the comparatively smooth unbroken surface of the plateau, which by the way is devoid of a central cone. The surrounding features and objects indicated in our illustration are of the highest interest, and a few of them demand special description.

THE VALLEY OF THE ALPS (37·0—86·0). PLATE XIV.

This remarkable object lays somewhat diagonally to the west of Plato; when seen with a low magnifying power, (80 or 100), it appears as a rut or groove tapering towards each extremity. It measures upwards of 75 miles long by about six miles wide at the broadest part. When examined under favourable circumstances, with a magnifying power of from 200 to 300, it is seen to be a vast flat-bottomed valley bordered by gigantic mountains, some of which attain heights upwards of 10,000 feet; towards the south-east of this remarkable valley, and on both sides of it, are groups of isolated mountains, several of which are fully 8000 feet high. This flat-bottomed valley, which has retained the integrity of its form amid such disturbing forces as its immediate surroundings indicate, is one of the many structural enigmas with which the lunar surface abounds. To the north-west of the valley a vast number of isolated mounds or small mountains of exudation may be seen; so numerous are they as to defy all attempts to count them with anything like exactness; and among them, a power of 200 to 300 will enable an observer, under favourable circumstances, to detect vast numbers of small but perfectly-formed craters.

PICO, 211 (41·9—87·3). PLATE XIV.

This is one of the most interesting examples of an isolated volcanic "mountain of exudation," and it forms a very striking object when seen under favourable circumstances. Its height is upwards of 8000 feet, and it is about three times as long at the base as it is broad. The summit is cleft into three peaks, as may be ascertained by the three-peaked shadow it casts on the plain. Five or six minute craters of very perfect

M

form may be detected close to the base of this magnificent mountain. There are several other isolated peaks or mountains of the same class within 30 or 40 miles of it which are well worthy of careful scrutiny, but Pico is the master of the situation, and offers a glorious subject for realizing a lunar day-dream in the mind's eye, if we can only by an effort of imagination conceive its aspect under the fiercely brilliant sunshine by which it is illuminated, contrasted with the intensely black lunar heavens studded with stars shining with a steady brightness of which, by reason of *our* atmosphere intervening, we can have no adequate conception save by the aid of a well-directed imagination.

TYCHO, 30 (43·0—142·3). PLATE XVI.

This magnificent crater, which occupies the centre of the crowded group in our Plate, is 54 miles in diameter, and upwards of 16,000 feet deep, from the highest ridge of the rampart to the surface of the plateau, whence rises a grand central cone 5000 feet high. It is one of the most conspicuous of all the lunar craters, not so much on account of its dimensions as from its occupying the great focus of disruption from whence diverge those remarkable bright streaks, many of which may be traced over 1000 miles of the moon's surface, disregarding in their course all interposing obstacles. There is every reason to conclude that Tycho is an instance of a vast disruptive action which rent the solid crust of the moon into radiating fissures, which were subsequently occupied by extruded molten matter, whose superior luminosity marks the course of the cracks in all directions from the crater as their common centre of divergence. So numerous are these bright streaks when examined by the aid of the telescope, and they give to this region of the moon's surface such an extra degree of luminosity, that, when viewed as a whole, their locality can be distinctly seen at full moon by the unassisted eye as a bright patch of light on the southern portion of the disc. (See Plate III.) The causative origin of the streaks is discussed and illustrated in Chapter XI.

The interior of this fine crater presents striking examples of the concentric terrace-like formations that we have elsewhere assigned to vast landslip actions. Somewhat similar concentric terraces may be observed in other lunar craters; some of these, however, appear to be the results of some temporary modification of the ejective force, which has caused the formation of more or less perfect inner ramparts : what we conceive to be true landslip terraces are always distinguished from these by their more or less fragmentary character.

On reference to Plate III., showing the full moon, a very remarkable and special appearance will be observed in a dingy district or zone immediately surrounding the exterior of the rampart of Tycho, and of which we venture to hazard what appears to us a rational explanation : namely, that as Tycho may be considered to have acted as a sort of safety-valve to the rending and ejective force which caused, in the first instance, the cracking of this vast portion of the moon's crust—the molten matter that appears to have been forced up through these cracks, on finding a comparatively free exit by the vent of Tycho, so relieved the district immediately around him as to have thereby reduced, in amount, the exit of the molten matter, and so left a zone comparatively free from the extruded lava which, according to our view of the subject, came up simultaneously through the innumerable fissures, and, spreading sideways along their courses, left everlasting records of the original positions of the radiating cracks in the form of the bright streaks which we now behold.

"WARGENTIN," 26 (57·5—140·2). PLATE XVII.

This object is quite unique of its kind—a crater about 53 miles across that to all appearance has been filled to the brim with lava that has been left to consolidate. There are evidences of the remains of a rampart, especially on the south-west portion of the rim. The general aspect of this extraordinary object has been not unaptly compared to a "thin cheese." The terraced and rutted exterior of the rampart has all the usual characteristic details of the true crater. The surface of the high plateau is marked by a few ridges branching from a point nearly in its centre, together with some other slight elevations and depressions ; these, however, can only be detected when the sun's rays fall nearly parallel to the surface of the plateau.

To the north of this interesting object is the magnificent ring formation Schickard, whose vast diameter of 123 miles contrasts strikingly with that of the sixteen small craters within his rampart, and equally so with a multitude of small craters scattered around. There are many objects of interest on the portion of the lunar surface included within our illustration, but as they are all of the usual type, we shall not fatigue the attention of our readers by special descriptions of them.

ARISTARCHUS, 176 (6·3—99·2), AND HERODOTUS, 175 (63·2—99·6). PLATE XVIII.

These two fine examples of lunar volcanic craters are conspicuously situated in the north-east quarter of the moon's disc. Aristarchus has a circular rampart nearly 28 miles diameter, the summit of which is about 7500 feet above the surface of the plateau, while its height above the general surface of the moon is 2600 feet. A central cone having several subordinate peaks completes the true volcanic character of this crater : its rampart banks, both outside and inside, have fine examples of the segmental crescent-shaped ridges or landslips, which form so constant and characteristic a feature in the structure of lunar craters. Several very notable cracks or chasms may be seen to the north of these two craters. They are contorted in a very unusual and remarkable manner, the result probably of the force which formed them having to encounter very varying resistance near the surface.

Some parts of these chasms gape to the width of two to three miles, and when closely scrutinized are seen to be here and there partly filled by masses which have fallen inward from their sides. Several smaller craters are scattered around, which, together with the great chasms and neighbouring ridges, give evidence of varied volcanic activity in this locality. We must not omit to draw attention to the parallelism or general similarity of "strike" in the ridges of extruded matter ; this appearance has special interest in the eyes of geologists, and is well represented in our illustration.

Aristarchus is specially remarkable for the extraordinary capability which the material forming its interior and rampart banks has of reflecting light. Although there are many portions of the lunar surface which possess the same property, yet few so remarkably as in the case of Aristarchus, which shines with such brightness, as compared with its immediate surroundings, as to attract the attention of the most unpractised observer. Some have supposed this appearance to be due to active volcanic discharge still lingering on the lunar surface, an idea in which, for reasons to be duly adduced, we have no faith. Copernicus, in the remarkable bright streaks which radiate from it, and Tycho also, as well as several other spots, are apparently composed of material very nearly as highly reflective as that of Aristarchus. But the comparative isolation of Aristarchus, as well as the extraordinary light-reflecting property of its material, renders it especially noticeable, so much so as to make it quite a conspicuous object when illuminated only by earth-light, when but a slender crescent of the lunar disc is illuminated, or when, as during a lunar eclipse, the disc of the moon is within the shadow of the earth and is lighted only by the rays refracted through the earth's atmosphere.

There are no features about Herodotus of any such speciality as to call for remark, except it be the breach of the north side of its rampart by the southern extremity of a very remarkable contorted crack or chasm, which to all appearance owes its existence to some great disruptive action subsequent to the formation of the crater.

WALTER, 48 (37·8—131·0), AND ADJACENT INTRUSIVE CRATERS. PLATE XX.

This Plate represents a southern portion of the moon's surface, measuring 170 by 230 miles. It includes upwards of 200 craters of all dimensions, from Walter, whose rampart measures nearly 70 miles across, down to those of such small apparent diameter as to require a well-practised eye to detect them. In the interior of the great crater Walter a remarkable group of small craters may be observed surrounding his central cone, which in this instance is not so perfectly in the centre of the rampart as is usually the case. The number of small craters which we have observed within the rampart is 20, exclusive of those on the rampart itself. The entire group represented in the Plate suggests in a striking manner the wild scenery which must characterize many portions of the lunar surface ; the more so if we keep in mind the vast proportions of the objects which they comprise, upon which point we may remark that the smallest crater represented in this Plate is considerably larger than that of Vesuvius.

ARCHIMEDES, 191 (40·3—95·8), AUTOLYCUS, 189 (36·8—95·5), ARISTILLUS, 190 (37·0—93·3), AND THE APPENNINES. PLATE IX.

This group of three magnificent craters, together with their remarkable surroundings, especially including the noble range of mountains termed the Apennines, forms on the whole one of the most striking and interesting portions of the lunar surface. If the reader is not acquainted with what the telescope can reveal as to the grandeur of the effect of sunrise on this very remarkable portion of the moon's surface, he should carefully inspect and study our illustration of it; and if he will pay due regard to our previously repeated suggestion concerning the attached scale of miles, he will, should he have the good fortune to study the actual objects by the aid of a telescope, be well prepared to realize and duly appreciate the magnificence of the scene which will be presented to his sight.

Were we to attempt an adequate detail description of all the interesting features comprised within our illustration, it would, of itself, fill a goodly volume; as there is included within the space represented every variety of feature which so interestingly characterizes the lunar surface. All the more prominent details of types of their class; and are so favourably situated in respect to almost direct vision, as to render their nature, forms, and altitudes above and depths below the average surface of the moon most distinctly and impressively cognizable.

Archimedes is the largest crater in the group; it has a diameter of upwards of 52 miles, measuring from summit to summit of its vast circular rampart or crater wall, the average height of which, above the plateau, is about 4300 feet; but some parts of it rise considerably higher, and, in consequence, cast steeple-like shadows across the plateau when the sun's rays are intercepted by them at a low angle. The plateau of this grand crater is devoid of the usual central cone. Two comparatively minute but beautifully-formed craters may be detected close to the north-east interior side of the surrounding wall of the great crater. Both outside and inside of the crater wall may be seen magnificent examples of the landslip subsidence of its overloaded banks; these landslips form vast concentric segments of the outer and inner circumference of the great circular rampart, and doubtless belong to its era of formation. Two very fine examples of cracks, or chasms, may be observed proceeding from the opposite external sides of the crater, and extending upwards of 100 miles in each direction ; these cracks, or chasms, are fully a mile wide at their commencement next the crater, and narrow away to invisibility at their further extremity. Their course is considerably crooked, and in some parts they are partially filled by masses of the material of their sides, which have fallen inward and partially choked them. The depths of these enormous chasms must be very great, as they probably owe their existence to some mighty upheaving action, which there is every reason to suppose originated at a profound depth, since the general surface on each side of the crater does not appear to be disturbed as to altitude, which would have been the case had the upheaving action been at a moderate depth beneath. We would venture to ascribe a depth of not less than ten miles as the most moderate estimate of the profundity of these terrible chasms. If the reader would realize the scale of them, let him for a moment imagine himself a traveller on the surface of the moon coming upon one of them, and finding his onward progress arrested by the sudden appearance of its vast black yawning depths ; for by reason of the angle of his vision being almost parallel to the surface, no appearance of so profound a chasm would break upon his sight until he came comparatively close to its fearful edge. Our imaginary lunar traveller would have to make a very long détour, ere he circumvented this terrible interruption to his progress. If the reader will only endeavour to realize in his mind's eye the terrific grandeur of a chasm a mile wide and of such dark profundity as to be, to all appearance, fathomless—portions of its rugged sides fallen in wild confusion into the jaws of the tortuous abyss, and catching here and there a ray of the sun sufficient only to render the darkness of the chasm more impressive as to its profundity—he will, by so doing, learn to appreciate the romantic grandeur of this, one of the many features which the study of the lunar surface presents to the careful observer, and which exceed in sublimity the wildest efforts of poetic and romantic imagination. The contemplation of these views of the lunar world are, moreover, vastly enhanced by especial circumstances which add greatly to the impressiveness of lunar scenery, such as the unchanging pitchy-black aspect of the heavens and the death-like silence which reigns unbroken there.

These digressions are, in some respects, a forestallment of what we have to say by-and-by, and so far they are out of place; but with the illustration to which the above remarks refer placed before the reader, they may, in some respects, enhance the interest of its examination.

The upper portion of our illustration is occupied by the magnificent range of volcanic mountains named after our Appennines, extending to a length of upwards of 450 miles. This mountain group rises gradually from a comparatively level surface towards the south-west, in the form of innumerable comparatively small mountains of exudation, which increase in number and altitude towards the north-east, where they culminate and suddenly terminate in a sublime range of peaks, whose altitude and rugged aspect must form one of the most terribly grand and romantic scenes which imagination can conceive. The north-east face of the range terminates abruptly in an almost vertical precipitous face, and over the plain beneath intense black steeple or spire-like shadows are cast, some of which at sunrise extend fully 90 miles, till they lose themselves in the general shading due to the curvature of the lunar surface. Nothing can exceed the sublimity of such a range of mountains, many of which rise to heights of 18,000 to 20,000 feet at one bound from the plane at their north-east base. The most favourable time to examine the details of this magnificent range is from about a day before first quarter to a day after, as it is then that the general structure of the range as well as the character of the contour of each member of the group can, from the circumstances of illumination then obtaining, be most distinctly inferred.

Several comparatively small perfectly-formed craters are seen interspersed among the mountains, giving evidence of the truly volcanic character of the surrounding region, which, as before said, comprises in a comparatively limited space the most perfect and striking examples of nearly every class of lunar volcanic phenomena.

We have endeavoured on Plate XXIII. to give some idea of a landscape view of a small portion of this mountain range.

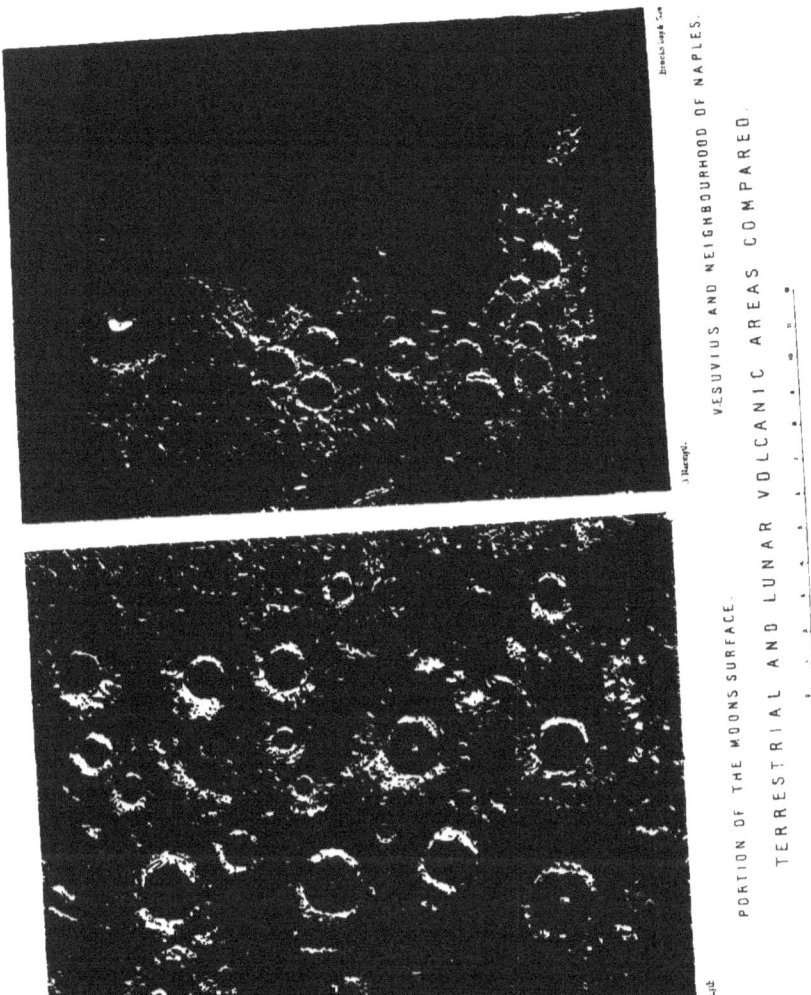

PLATE VI

PORTION OF THE MOON'S SURFACE.

VESUVIUS AND NEIGHBOURHOOD OF NAPLES.

TERRESTRIAL AND LUNAR VOLCANIC AREAS COMPARED.

CHAPTER VIII.

ON LUNAR CRATERS.

As we stated in our brief general description of the visible hemisphere of the moon, and as a cursory glance at our map and plates will have shown, the predominant features of the lunar surface are the circular or amphitheatrical formations that, by their number, and from their almost unnatural uniformity of design, induced the belief among early observers that they must have been of artificial origin. In proceeding now to examine the details of our subject with more minuteness than before, these angular formations claim the first share of our attention.

By general acceptation the term "crater" has been used to represent nearly all the circular hollows that we observe upon the moon; and without doubt the word in its literal sense, as indicating a *cup* or circular cavity, is so far aptly applied. But among geologists it has been employed in a more special sense to define the hollowing out that is found at the summit of some extinct, and the majority of active, volcanoes. In this special sense it may be used by the student of the lunar surface, though in some, and indeed in the majority of cases, the lunar crater differs materially in its form with respect to its surroundings from those on the earth; for while, as we have said, the terrestrial crater is generally a hollow on a mountain top with its flat bottom high above the level of the surrounding country, those upon the moon have their lowest points depressed more or less deeply below the general surface of the moon, the external height being frequently only a half or one-third of the internal depth. Yet are the lunar craters truly volcanic; as Sir John Herschel has said, they offer the true volcanic character *in its highest perfection*. We have upon the earth some few instances in which the geological conditions which have determined the surface-formation have been identical with those that have obtained upon the moon; and as a result we have some terrestrial volcanic districts that, could we view them under the same circumstances, would be identical in character with what we see by telescopic aid upon our satellite. The most remarkable case of this similarity is offered by a certain tract of the volcanic area about Naples, known from classic times as the *Campi Phlegræi*, or burning fields, a name given to them in early days, either because they showed traces of ancient earth-fire, or because there were attached to the localities traditions concerning hot-springs and sulphurous exhalations, if not of actual fiery eruptions. The resemblance of which we are speaking is here so close that Professor Phillips, in his work on Vesuvius, which by the way contains a historical description of the district in question, calls the moon a grand Phlegreian field. How closely the ancient craters of this famous spot resemble the generality of those upon the moon may be judged from Plate VI., in which representations of two areas, terrestrial and lunar, of the same extent, are exhibited side by side, the terrestrial region being the volcanic neighbourhoood of Naples, and the lunar a portion of the surface about the crater Theophilus.

In comparing these volcanic circles together, we are however brought face to face with a striking difference that exists between the lunar and terrestrial craters. This is the difference of magnitude. None of those Plutonian amphitheatres included in the terrestrial area depicted exceed a mile in diameter, and few larger volcanic vents than these are known upon the earth. Yet when we turn to the moon, and measure some of the larger craters there, we are astonished to find them ranging from an almost invisible minuteness to 74 miles in diameter. The same disproportion exists between the depths of the two classes of craters. To give an idea of relative dimensions, we would refer to our illustration of Copernicus* and its hundreds of comparatively minute surrounding craters. Our terrestrial Vesuvius would be represented by one of these last, which upon the plate measures about the twentieth of an inch in diameter! And this disproportion strikes us the more forcibly when

* Plate VIII.

we consider that the lunar globe has an area only one-thirteenth of that of the earth. In view of this great apparent discrepancy it is not surprising that many should have been incredulous as to the true volcanic character of the lunar mountains, and have preferred to designate them by some "non-committal" term, as an American geologist (Professor Dana) has expressed it. But there is a feature in the majority of the ring-mountains that, as we conceive, demonstrates completely the fact of volcanic force having been in full action, and that seems to stamp the volcanic character upon the crater-forms. This special feature is the central cone, so well known as a characteristic of terrestrial volcanoes, accepted as the result of the last expiring effort of the eruptive force, and formed by the deposit, immediately around the volcanic orifice, of matter which there was not force enough to project to a greater distance. Upon the moon we have the central cone in small craters comparable to those on the earth, and we have it in progressively larger examples, upon all scales, up to craters of 74 miles in diameter, as we have shown in Plate VII. Where, then, can we draw the line ? Where can we say the parallel action to that which placed Vesuvius in or near the centre of the arc of Somma, or the cone figured in our sectional drawing of Vesuvius (Fig. 3) in the middle of its present crater—where can we say that the action in question ceased to manifest itself on the moon, seeing that there is no break in the continuity of the crater-and-cone system upon the moon anywhere between craters of $1\frac{3}{4}$ miles and 74 miles in diameter? We have, it is true, many examples of coneless craters, but these are of all sizes, down to the smallest, and up to a largeness that *would* almost seem to render untenable the ejective explanation : of these we shall specially speak in turn, but for the present we will confine ourselves to the normal class of lunar craters, those that have central cones, and that are in all reasonable probability truly volcanic.

And in the first place let us take a passing glance at the probable formative process of a terrestrial volcano. Rejecting the hypothesis of Von Buch, which geologists have on the whole found to be untenable, and which ascribes the formation of all mountains to the elevation of the earth's crust by some thrusting power beneath, we are led to regard a volcano as a pyramid of ejected matter, thrown out of and around an orifice in the external solid shell of the earth by commotions engendered in its molten nucleus. What is the precise nature and source of the ejective force geologists have not perfectly agreed upon, but we may conceive that highly expanded vapour, in all probability steam, is its primary cause. The escaping aperture may have been a weak place since the foundations of the earth were laid, or it may have been formed by a local expansion of the nucleus in the act of cooling, upon the principle enunciated in our Third Chapter; or, again, the expansile vapour may have forced its own way through that point of the confining shell that offered it the least resistance. The vent once formed, the building of the volcanic mountain commenced by the out-belching of the lava, ashes, and scoria, and the dispersion of these around the vent at distances depending upon the energy with which they were projected. As the action continued, the ejected matter would accumulate in the form of a mound, through the centre of which communication would be maintained with the source of the ejected materials and the seat of the explosive agency. The height to which the pile would rise must depend upon several conditions: upon the steady sustenance of the matter, and upon the form and weight of the component masses, which will determine the slope of the mountain's sides. Supposing the action to subside gradually, the tapering form will be continued upwards by the comparatively gentle deposition of material around the orifice, and a perfect cone will result of some such form as that represented below, which is the outline ascribed by

FIG. 16.

Professor Phillips to Vesuvius in pre-historic, or even pre-traditional times, and which may be seen in its full integrity in the cases of Etna, Teneriffe, Fussi-Yamma, the great volcanic mountain of Japan, and many others. The earliest recorded form of Vesuvius is that of a truncated cone represented in Fig. 17, which shows its condition, according to

PLATE VII

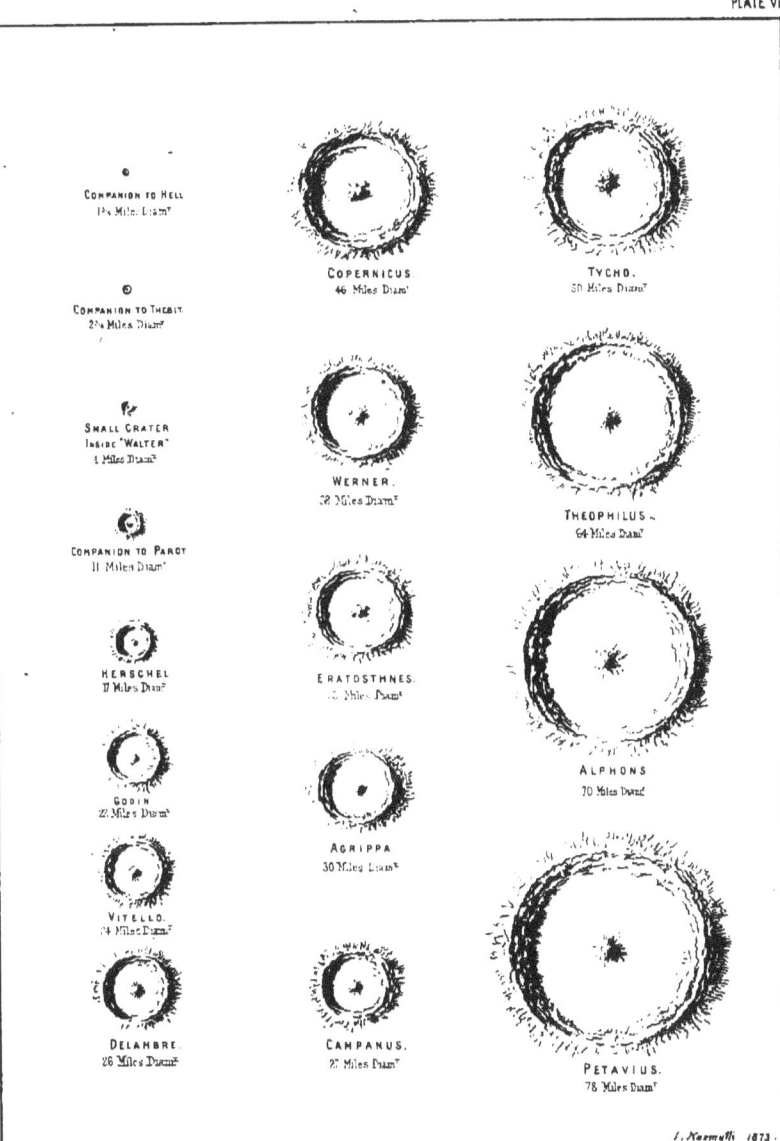

DIAGRAM OF LUNAR CRATERS FORMING A SERIES RANGING FROM 1¾ MILES TO 78 MILES DIAMETER, ALL CONTAINING CENTRAL CONES.

Published by John Murray, Albemarle Street Piccadilly

Strabo, in the century preceding the Christian Era. Now this form may have been assumed under two conditions. If, as Phillips has surmised, the mountain originally had a peaked summit with but a small

FIG. 17.

crater-orifice, at the point, then we must ascribe its decapitation to a subsequent eruption which in its violence carried away the upper portion, either suddenly, or through a comparatively slow process of grinding away or widening out of the sides of the orifice by the chafing or fluxing action of the out-going materials. But it is probable that the mountain never had the perfect summit indicated in our first outline. The violent outburst that caused the great crater-opening of our second figure may have been but one paroxysmal phase of the eruption that built the mountain: a sudden cessation of the eruptive force when at its greatest intensity, and when the orifice was at its widest, would leave matters in an opposite condition to that suggested as the result of a slow dying out of the action: instead of the peak we should have a wide crater-mouth. It is of small consequence for our present purpose whether the crater was contemporaneous with the primitive formation of the mountain, or whether it was formed centuries afterwards by the blowing away of the mountain's head; for upon the vast scale of geological time, intervals such as those between successive paroxysms of the same eruption, and those between successive eruptions, are scarcely to be discriminated, even though the first be days and the second centuries. We may remark that the widening of a crater by a subsequent and probably more powerful

FIG. 18.

eruption than that which originally produced it is well established. We have only to glance at the sketch, Fig. 18, of the outline of Vesuvius as it appeared between the years A.D. 79 and 1631 to see how the old crater was enlarged by the terrible Pompeian eruption of the first-mentioned year. Here we have a crater ground and blown away till its original diameter of a mile and three-quarters has been increased to nearly three miles. Scrope had no hesitation in expressing his conviction that the external rings, such as those of Santorin, St. Jago, St. Helena, the Cirque of Teneriffe, the Curral of Madeira, the cliff range that surrounds the island of Bourbon, and others of similar form and structure, however wide the area they enclose, are truly the " basal wrecks" of volcanic mountains that have been blown into the air each by some eruption of peculiar paroxysmal violence and persistence; and that the circular or elliptical basins which they wholly or in part surround are in all cases true craters of eruption.

When the violent outburst that produces a great crater in a volcanic mountain-top more or less completely subsides, the funnel or escaping orifice becomes choked with débris. Still the vent strives to keep itself open, and now and then gives out a small delivery of cindery matter, which, being piled around the vent, after the manner of its great prototype, forms the inner cone. This last may in its turn bear an open crater upon its summit, and a still smaller cone may form within it. As the action further dies away, the molten lava, no longer seething and boiling, and spirting forth with the rest of the ejected matter, wells upwards slowly, and cooling rapidly as it comes in contact with the atmosphere, solidifies and forms a flat bottom or floor to the crater.

It may happen that a subsequent eruption from the original vent will be comparable in violence to the original one, and then the inner cone assumes a magnitude that renders it the principal feature of the mountain, and reduces the old crater to a secondary object. This has been the case with Vesuvius. During the eruption of 1631 the great cone which we now call Vesuvius was thrown up, and the ancient crater now distinguished as Monte Somma became a subsidiary portion of the whole mountain.

FIG. 19.

Then the appearance was that shown in Fig. 19, and which does not differ greatly from that presented in the present day. The summit of the Vesuvian cone, however, has been variously altered; it has been blown away, leaving a large crateral hollow, and it has rebuilt itself nearly upon its former model.

When we transfer our attention to the volcanoes of the moon, we find ourselves not quite so well favoured with means for studying the process of their formation; for the sight of the building up of a volcanic mountain such as man has been permitted to behold upon the earth has not been allowed to an observer of the moon. The volcanic activity, enfeebled though it now be, of which we are witnesses from time to time on the earth, has altogether ceased upon our satellite, and left us only its effects as a clue to the means by which they were produced. If we in our time could have seen the actual throwing up of a lunar crater, our task of description would have been simple; as it is we are compelled to infer the constructive action from scrutiny of the finished structure.

We can scarcely doubt that where a lunar crater bears general resemblance to a terrestrial crater, the process of formation has been nearly the same in the one case as in the other. Where variations present themselves they may reasonably be ascribed to the difference of conditions pertaining to the two spheres. The greatest dissimilarity is in the point of dimensions; the projection of materials to 20 or more miles distance from a volcanic vent appears almost incredible, until we realize the full effect of the conditions which upon the moon are so favourable to the dispersive action of an eruptive force. In the first place, the force of gravity upon our satellite is only one-sixth of that to which bodies are subject upon the earth. Secondly, by reason of the small magnitude of the moon and its proportionally much larger surface in ratio to its magnitude, the rate at which it parted with its cosmical heat must have been much more rapid than in the case of the earth, especially when enhanced by the absence of the heat-conserving power of an atmosphere of air or water vapour; and the disruptive and eruptive action and energy may be assumed to be greater in proportion to the more rapid rate of cooling; operating, too, as eruptive action would on matter so much reduced in weight as it is on the surface of the moon, we thus find in combination conditions most favourable to the display of volcanic action in the highest degree of violence. Moreover, as the ejected material in its passage from the centre of discharge had not to encounter any atmospheric resistance, it was left free to continue the primary impulse of its ejection without other than gravitative diminution, and thus to deposit itself at distances from its source vastly greater than those of which we have examples on the earth.

We can of course only conjecture the source or nature of the moon's volcanic force. If geologists have had difficulty in assigning an origin to the power that threw up our earthly volcanoes, into whose craters they can penetrate, whose processes they can watch, and whose material they can analyze, how vastly more difficult must be the inquiry into the primary source of the power that has been at work upon the moon, which cannot be virtually approached by the eye within a distance of six or eight hundred miles, and the material of which we cannot handle to see if it be compacted by heat, or distended by vapours. Steam is the agent to which geologists have been accustomed to look for explanation of terrestrial volcanoes; the contact of water with the molten nucleus of our globe is accepted as a probable means whereby volcanic commotions are set up and ejective action is generated. But we are debarred from referring to steam as an element of lunar geology, by reason of the absence of water from the lunar globe. We might suppose that a small proportion of water once existed; but a small proportion would not account for the immense display of volcanic action which the whole surface exhibits. If we admitted a Neptunian origin to the disturbances of the moon's crust, we should be compelled to suppose that water had existed nearly in as great quantity, area for area, there as upon our globe; but this we cannot reasonably do.

Aqueous vapour being denied us, we must look in other directions for an ejective force. Of the nature of the lunar materials we can know nothing, and we might therefore assume anything; some have had recourse to the supposition of expansive vapours given off by some volatile component of the said material while in a state of fusion, or generated by chemical combinations. Professor Dana refers to sulphur as probably an important element in the moon's geology, suggesting this substance because of the part which it appears to play in the volcanic or igneous

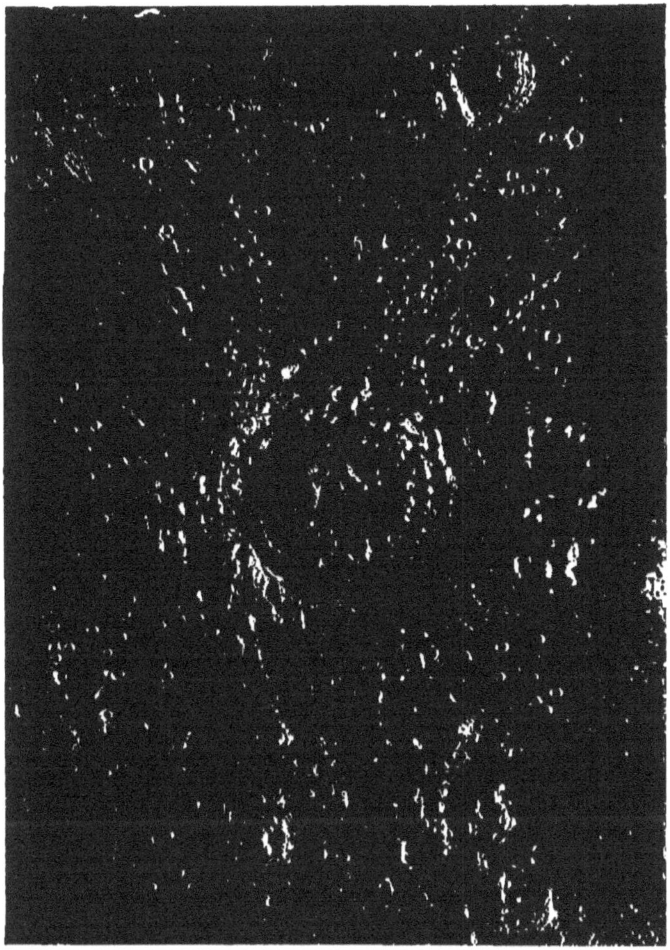

PLATE VIII

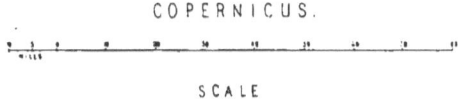

COPERNICUS.

SCALE

Published by John Murray Albemarle Street Piccadilly

operations of our globe, and on account of its presence in cosmical meteors that have come within range of our analysis. Any matter sublimated by heat in the substrata of the moon would be condensed upon reaching the cold surrounding space, and would be deposited in a state of fine powder, or otherwise in a solid form. Maedler has attributed the highly reflective portions of some parts of the surface, such as the bright streams that radiate from some of the craters, Copernicus and Tycho for instance, ✸ the vitrification of the surface matter by gaseous currents. But in suppositions like these we must remember that the probability of truth diminishes as the free ground for speculation widens. It does not appear clear how expansive vapours could have lain dormant till the moon assumed a solid crust, as all such would doubtless make their escape before any shell was formed, and at an epoch when there was ample facility for their expansion.

While we are not insensible of the value of an expansive vapour explanation, if it could be based on anything beyond mere conjecture, we are disposed to attach greater weight to that afforded by the principle sketched in our third chapter, viz., of expansion upon solidification. We gave, as we think, ample proof that molten matter of volcanic nature, when about passing to the solid state, increases its bulk to a considerable degree, and we suggested that the lunar globe at one period of its history must have been, what our earth is now, a solid shell encompassing a molten nucleus; and further, that this last, in approaching its solid condition, expanded and burst open or rent its confining crust. At first sight it may seem that we are ascribing too great a degree of energy to the expansive force which molten substances exhibit in passing to the solid condition, seeing that in general such forces are slow and gradual in their action; but this anomaly disappears when we consider the vast bulk of the so expanding matter, and the comparatively small amount that in its expansion it had to displace. It is true that there are individual mountains on the moon covering many square miles of surface, that as much as a thousand cubic miles of material may have been thrown up at a single eruption; but what is this compared to the entire bulk of the moon itself? A grain of mustard-seed upon a globe three feet in diameter represents the scale of the loftiest of terrestrial mountains; a similar grain upon a globe one foot in diameter, would indicate the proportion of the largest upon the moon. A model of our satellite with the elevations to scale would show nothing more than a little roughness, or superficial blistering. Turn for a moment to our map (Plate IV.), upon which the shadows give information as to the heights of the various irregularities, and suppose it to represent the actual size of some sphere whose surface has been broken up by reactions of some kind of the interior upon the exterior—suppose it to have been a globe of fragile material filled with some viscous substance, and that this has expanded, cracked its shell, oozed out in the process of solidification, and solidified : the irregularity of surface which the small sphere, roughened by the out-leaking matter, would present, would not be less than that exhibited in the map under notice. When we say that a lunar crater has a diameter of 30 miles, we raise astonishment that such a structure could result from an eruption by the expansive force of solidifying matter ; but when we reflect that this diameter is less than the two-hundredth part of the circumference of the moon, we need have no difficulty in regarding the upheaval as the result of a force slight in comparison to the bulk of the material giving rise to it. We have upon the moon evidence of volcanic eruptions being the final result of most extensive dislocations of surface, such as could only be produced by some widely diffused uplifting force. We allude to the frequent occurrence of chains of craters lying in a nearly straight line, and of craters situated at the converging point of visible lines of surface disturbance. Our map will exhibit many examples of both cases. An examination of the upper portion (the southern hemisphere of the moon) will reveal abundant instances of the linear arrangement, three, four, five or even more crateral circles will be found to lie with their centres upon the same great-circle track, proving almost undoubtedly a connexion between them so far as the original disturbing force which produced them is concerned. Again, in the craters Tycho (30), Copernicus (147), Kepler (146), and Proclus (162), we see instances of the situation of a volcanic outburst at an obvious focus of disturbance. These manifest an up-thrusting force covering a large sub-surface area, and escaping at the point of least resistance. Such an extent of action almost precludes the gaseous explanation, but it is compatible with the expansion on consolidation theory, since it is reasonable to suppose that in the process of consolidation the viscous nucleus would manifest its increase of bulk over considerable areas, disturbing the superimposed crust either in one long crack, out of the wider opening parts of which the expanded material would find its escape, or "starring" it with numerous cracks, from the converging point of which the confined matter would be ejected in greatest abundance and, if ejected there with great energy and violence, would result in the formation of a volcanic crater.

The actual process by which a lunar crater would be formed would differ from that pertaining to a terrestrial crater only to the extent of the different conditions of the two globes. We can scarely accept Scrope's term "basal wrecks" (of volcanic mountains that have had the summits blown away) as applicable to the craters of the moon, for the reason that the lunar globe does not offer us any instance of a mountain comparable in extent to the great craters and whose summit has *not* been blown away. Scrope's definition implies a double, or divided process of formation : first the building up of a vast conical hill and then the decapitation and "evisceration" of it at some later period. There are grounds for this inferred double action among the terrestrial volcanoes, since both the perfect cone and its summitless counterpart are numerously exemplified. But upon the moon we have no perfect cone of great size, we have no exception whereby the rule can be proved. It is against probability, supposing every lunar crater to have once been a mountain, that in every case the mountain's summit should have been blown away ; and we are therefore compelled to consider that the moon's volcanic craters were formed by one continuous outburst, and that their "evisceration" was a part of the original formative process. We do not, however, include the central cone in this consideration : that may be reasonably ascribed to a secondary action or perhaps, better, to a weaker or modified phase of the original and only eruption.

Under these circumstances we conceive the upcasting and excavating of a normal lunar crater to have been primarily caused by a local manifestation of the force of expansion upon solidification of the subsurface matter of the moon, resulting in the creation of a mere "star" or crack in and through the outermost and solid crust. As we shall have to rely upon diagrams to explain the more complicated features, we give one of this elementary stage also as a commencement of the series; and Fig. 20 therefore represents a probable section of the lunar surface at a point

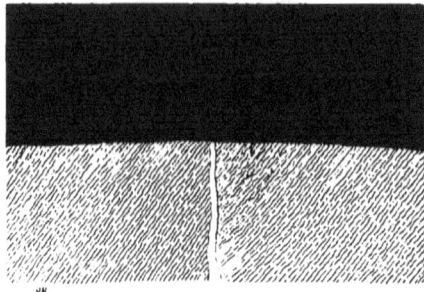

Fig. 20.

which was subsequently the location of a crater. From the vent thus formed we conceive the pent-up matter to have found its escape, not necessarily at a single outburst, but in all probability in a paroxysmal manner, as volcanic action manifests itself on our globe. The first outflow

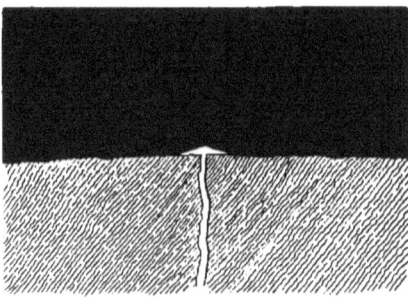

Fig. 21.

of molten material would probably produce no more than a mere hill or tumescence as shewn sectionally in Fig. 21; and if the ejective force

PLATE IX.

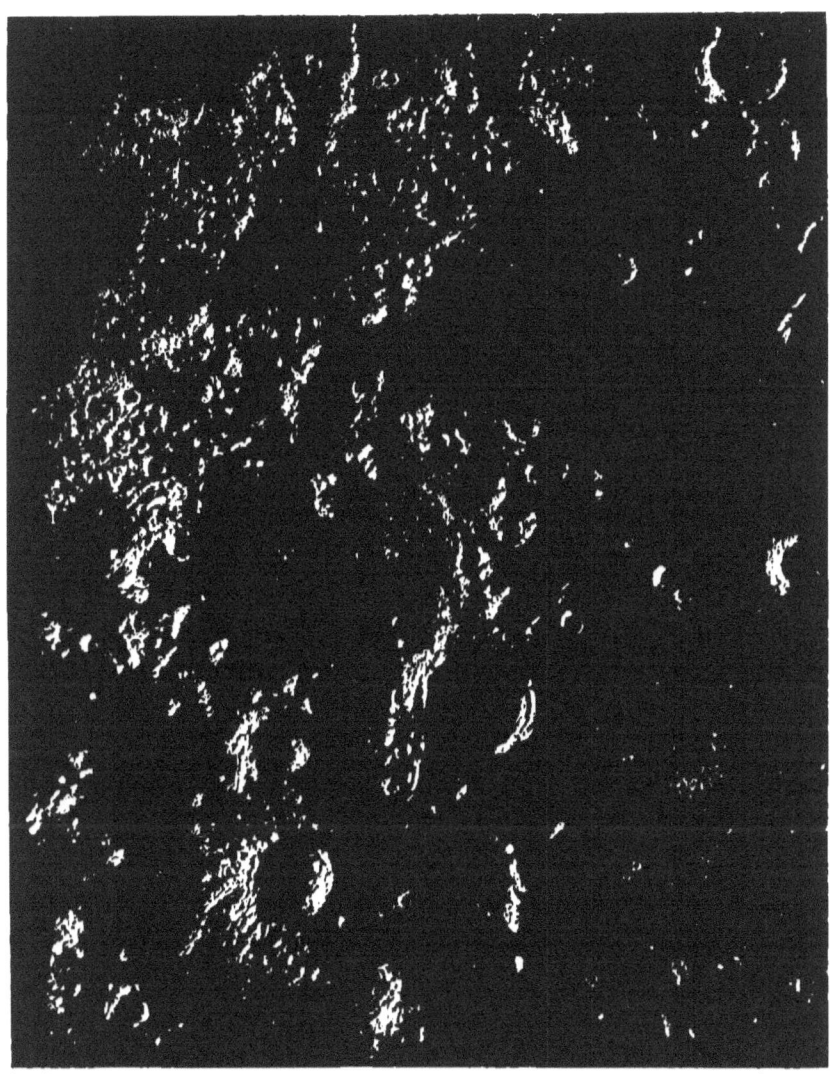

THE LUNAR APENNINES, ARCHEMIDES &c., &c.

SCALE.

were small this might increase to the magnitude of a mountain by an exudative process to be alluded to hereafter. But if the ejective force were violent, either at the moment of the first outburst or at any subsequent

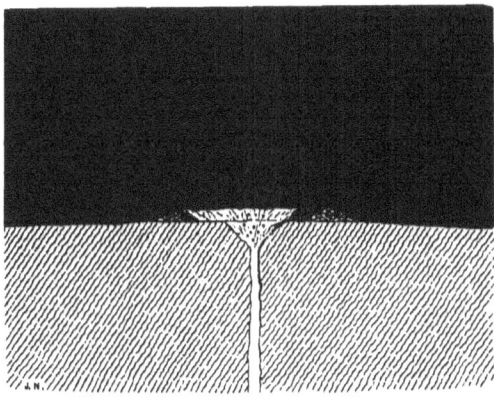

Fig. 22.

paroxysm, an action represented in Fig. 22 would result: the unsupported edges or lips of the vent-hole would be blown and ground or fluxed away, and a funnel-formed cavity would be produced, the ejected matter (so much of it as in falling was not caught by the funnel) being deposited around the hollow and forming an embryo circular mountain. The continuance of this action would be accompanied by an enlargement of the conical cavity or crater, not only by the outward rush of the violently discharged material, but also by the "sweating" or grinding action of such of it as in descending fell within the hollow. And at the same time that the crater enlarged the rampart would extend its circumference, for it would be formed of such material as did not fall back again into the crater. Upon this view of the crater-forming process we base the sketch, Fig. 23, of the probable section of a lunar crater at one period of its development.

So long as each succeeding paroxysm was greater than its predecessor, this excavating of the hollow and widening of its mouth and mound would be extended. But when a weaker outburst came, or when the energy of the last eruption died away, a process of slow piling up of

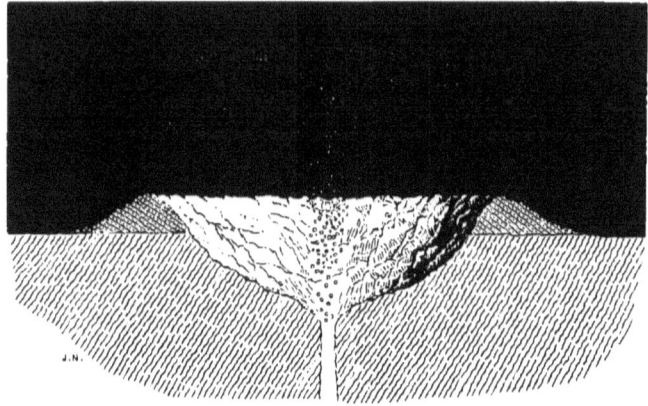

Fig. 23.

matter close around the vent would ensue. It is obvious that when the ejective force could no longer exert itself to a great distance it must merely have lifted its burden to the relieving vent and dropped it in the immediate neighbourhood. Even if the force were considerable, the effect, so long as it was insufficient to throw the ejecta beyond the rim of the crater, would be to pile material in the lowermost part of the cavity; for what was not cast over the edge would roll or flow down the inner slope and accumulate at the bottom. And as the eruption died away, it would add little by little to the heap, each expiring effort leaving the out-giving matter nearer the orifice, and thus building up the central cone that is so conspicuous a feature in terrestrial volcanoes, and which is also a marked one in a very large proportion of the craters of the moon. This formation of the cone is pictorially described by Fig. 24.

In the volcanoes of the earth we observe another action either concurrent with or immediately subsequent to the erection or formation of the cone: this is the outflow or the welling forth of fluid lava, which

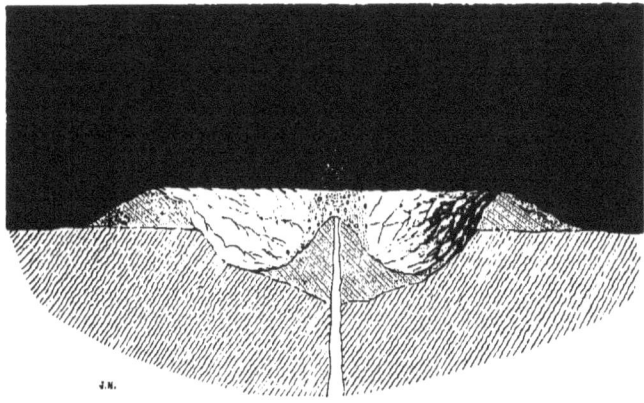

Fig. 24.

in cooling forms the well-known plateau. We have this feature copiously represented upon the moon and it is presumable that it has in general been produced in a manner analogous to its counterparts upon the earth. We may conceive that the fluid matter was either spirted forth with the solid or semisolid constituents of the cone, in which case it would drain down and fill the bottom of the crater; or we may suppose that it issued from the summit of the cone and ran down its sides, or that, as we see upon the earth, it found its escape before reaching the apex, by forcing its way through the basal parts. These actions are indicated hypothetically for the moon in Fig. 25; and the parallel phenomena for the earth are shewn by the actual case (represented in Fig. 26 and on Plate I.) of Vesuvius as it was seen by one of the authors in 1865, when the principal cone was vomiting forth ashes, stones, and red-hot lava, while a vent at the side emitted very fluid lava which was settling down and forming the plateau.

Although we cannot, obviously, see upon the moon evidence of a cone actually overtopped by the rising lake of lava, yet it is not unreasonable to suppose that such a condition of things actually occurred

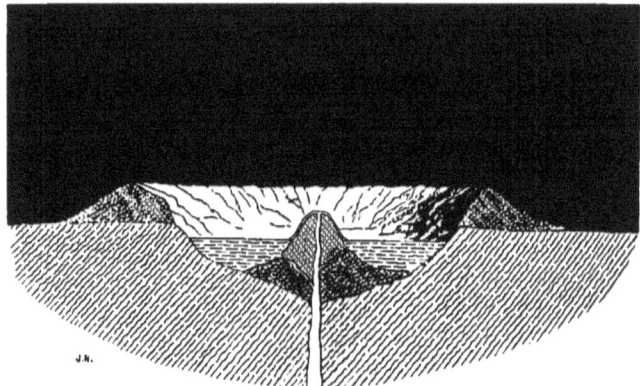

Fig. 25.

in many of those instances in which we observe craters without central cones, but with plateaux so smooth as to indicate previous fluidity or viscosity. From the state of things exhibited in Fig. 25 the transition to that shewn in Fig. 27 is easily, and to our view reasonably, conceivable. We are in a manner led up to this idea by a review of the various heights of central cones above their surrounding plateaux. For instance, in such examples as Tycho or Theophilus, we have cones high above the lava floor; in Copernicus, Arzachael and Alphonsus they are comparatively lower; the lava in these and some other craters does not appear to have risen so high; while in Aristotle and Eudoxus among others, we have only traces of cones, and it is supposable that in these cases the lava rose so high as nearly to overtop the central cones. Why should it not have risen so far as to overtop and therefore conceal some cones entirely? We offer this as at least a feasible explanation of some coneless craters: it is not necessary to suppose that it applies to

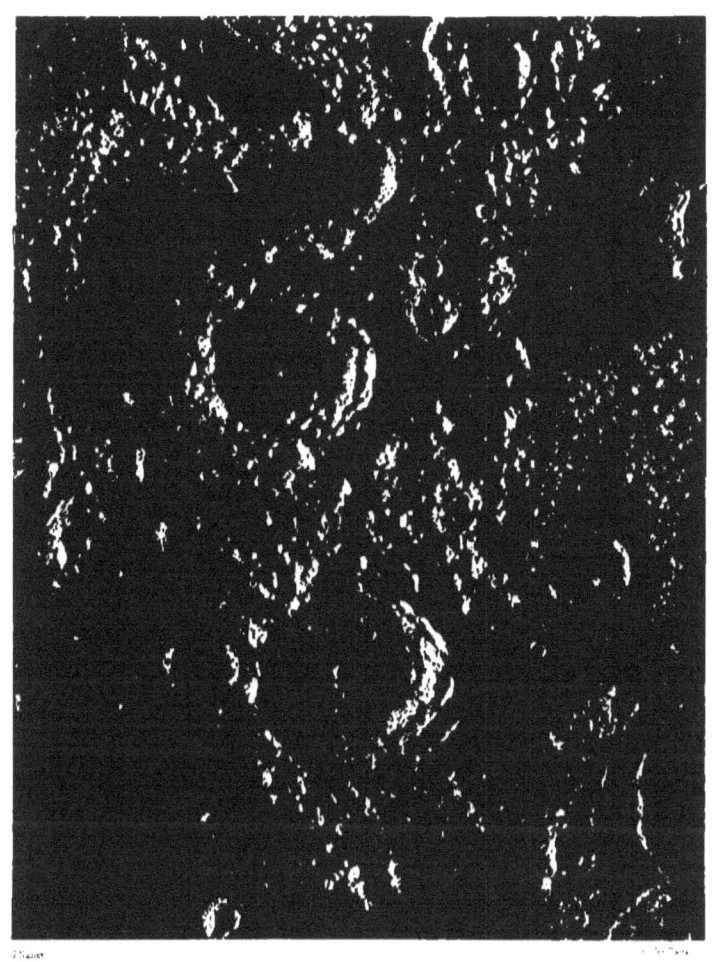

ARISTOTLE & EUDOXUS.

SCALE

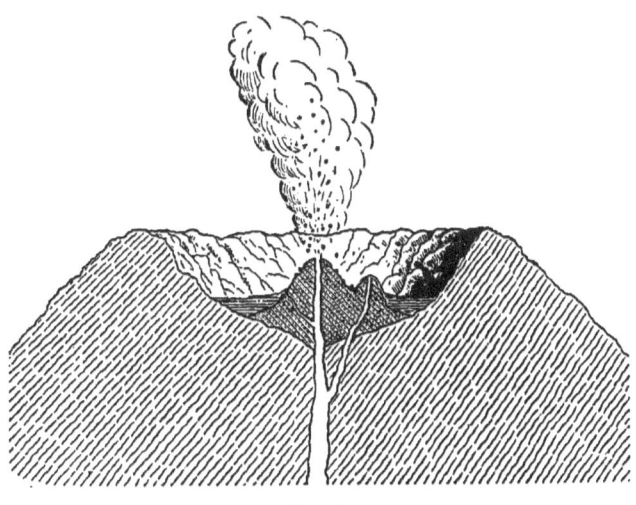

FIG. 26.

all such, however: there may have been many craters, the formation of

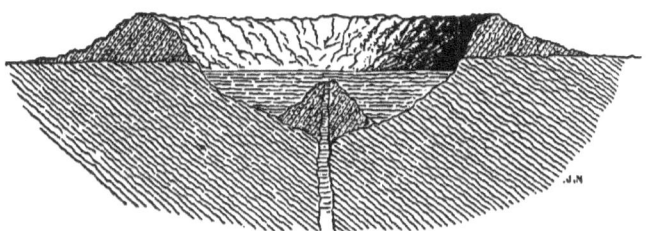

FIG. 27.

which ceased so abruptly that no cone was produced, though the welling forth of lava occurred from the vent, which may have been left fully open,

P

as in Fig. 28, or so far choked as to stay the egress of solid ejecta and yet allow the fluid material to ooze upwards through it, and so form a lake of molten lava which on consolidation became the plateau. As most of the examples of concless craters exhibit on careful examination minute craters on the surface of the otherwise smooth plateaux, we may suppose that such minute craters are evidences of the upflow of lava which resulted in the plateaux.

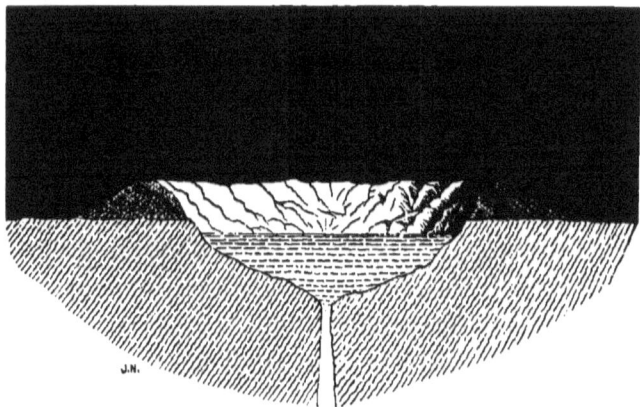

Fig. 28.

We have strong evidence in support of this up-flow of lava offered by the case of the crater Wargentin, (No. 26, 57·5—140·2) situated near the south-east border of the disc, and of which we give a special plate. (Plate XVII.) It appears to be really a crater in which the lava has risen almost to the point of overflowing, for the plateau is nearly level with the edge of the rampart. This edge appears to have been higher on one side than the other, for on the portion nearest the centre of the visible disc we may, under favourable circumstances, detect a segment of the basin's brim rising above the smooth plateau as indicated in our illustration. Upon the opposite side there is no such feature visible, the plateau forms a sharp table-like edge. It is just possible that an actual overflow of lava took place at this part of the crater, but from the unfavourable situation of this remarkable object it is impossible to decide the point by observation. There is no other crater upon the visible hemisphere of the moon that exhibits this filled-up condition ; but, unique as it is, it is sufficient to justify our conclusion that the plateau-forming action upon the moon has been a flowing-up of fluid matter from below subsequent to the formation of the crater-rampart, and not, as a casual glance at the great

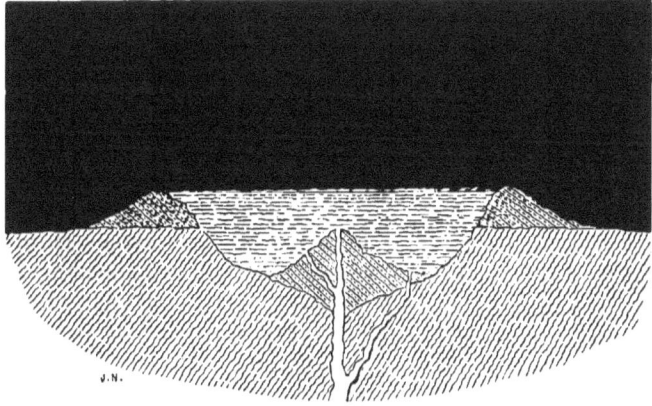

FIG. 29.

smooth-bottom craters might lead us to suspect, a result of some sort of diluvial deposit which has filled hollows and cavities and so brought up an even surface. The elevated basin of Wargentin could not have been filled thus while the surrounding craters with ramparts equally or less high remained empty : its contained matter must have been supplied from within, we must conjecture by the upflow of lava from the orifice which gave forth the material to form the crateral rampart in the first instance. We are free to conjecture that at some period of this table-mountain's formation it was a crater with a central cone, and that the rising lava over-topped this last feature in the manner shewn by the above figure (Fig. 29).

The question occurs whether other craters may not have been similarly filled and have emptied themselves by the bursting of the wall under the pressure of the accumulated lake of lava within. We know that this breaching is a common phenomenon in the volcanoes of our globe; the district of Auvergne furnishing us with many examples; and there are some suspicious instances upon the moon. Copernicus exhibits signs of such disruption, as also does the smaller crater intruding upon the great circle of Gassendi. (See Frontispiece.) But the existence of such discharging breaches implies the outpouring of a body of fluid or semi-fluid material, comparable in cubical content to the capacity of the crater, and of this we ought to see traces or evidence in the form of a bulky or extensive lava stream issuing from the breach. But although there are faint indications of once viscous material lying in the direction that escaping fluid would take, we do not find anything of the extent that we should expect from the mass of matter that would constitute a *craterfull*. It is true that if the escaping fluid had been very limpid it might have spread over a large area and have formed a stratum too thin to be detected. Such a degree of limpidity as would be required to fulfil this condition we are hardly, however, justified in assuming.

To return to the subject of central cones. Although there are cases in which the simple condition of a single cone exists, yet in the majority we see that the cone-forming process has been divided or interrupted, the consequence being the production of a group of conical hills instead of a single one. Copernicus offers an example of this character, six, some observers say seven, separate points of light, indicating as many peaks tipped with sunshine, having been seen when the greater part of the crater has been buried in shadow. Eratosthenes, Bulialdus, Maurolicus, Petavius, Langreen, and Gassendi, are a few among many instances of craters possessing more than a central single cone. This multiplication of peaks upon the moon doubtless arose from similar causes to those which produce the same feature in terrestrial volcanoes. Our sketch of Vesuvius in 1865 (Fig. 26) shows the double cone and the probable source of the secondary one in the diverted channel of the out-coming material. A very slight interruption in the first instance would suffice to divert the stream and form another centre of action, or a choking of the

TRIESNECKER.

SCALE

original vent would compel the issuing matter to find a less resisting thoroughfare into open space, and the process of cone-building would be continued from the new orifice, perhaps to be again interrupted after a time and again driven in another direction. In this manner, by repeated arrests and diversions of the ejecta, cone has grown upon the side of cone, till, ere the force has entirely spent itself, a cluster of peaks has been produced. It may have been that this action has taken place after the formation of the plateau, in the manner indicated by Fig. 30;

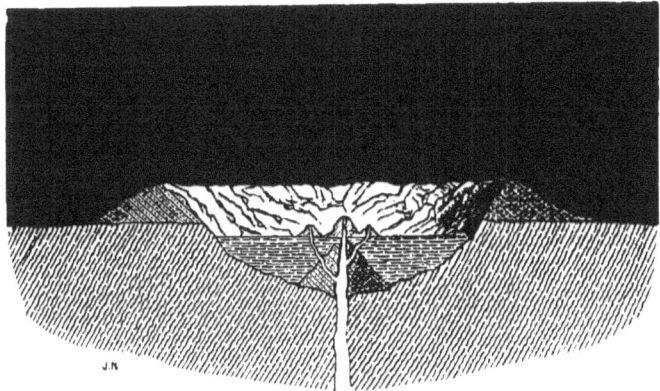

Fig. 30.

a spasmodic outburst of comparatively slight violence having sought relief from the original vent, and the flowing matter, finding the one orifice not sufficiently open to let it pass, having forced other exit through the plateau.

In frequent instances we observe the state of things represented in Fig. 31, in which the plateau is studded with few or many small craters. This is the case with Plato, with Arzachael, Hipparchus, Clavius (which contains about 15 small internal craters), and many others. It is probable that these subsidiary craters were produced by an after-action like that which has produced duplicated cones, but in which the secondary eruption has been of somewhat violent character, for it may almost be regarded as an axiom that violent eruptions excavate craters and weak ones pile up cones. In the cases referred to it seems reasonable to suppose that the main vent has been the channel for an up-cast of material, but that at some depth below the surface this material met with some obstruction or cause of diversion, and that it took a course which brought it out far away from the cone upon the floor of the plateau. It might even be carried so far as to be upon the rampart, and

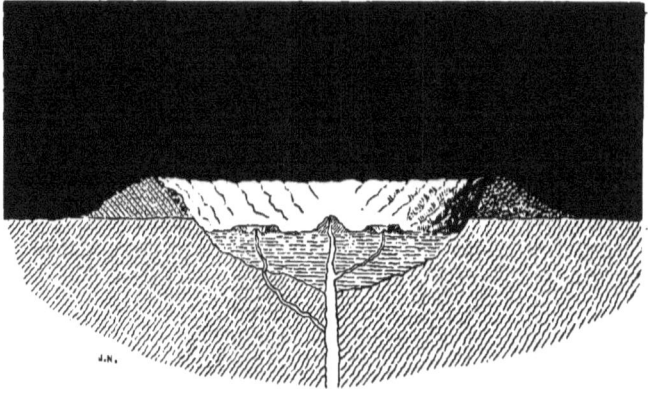

Fig. 31.

it is no uncommon thing to see small craters in such a situation, though when they appear at such a distance from the primary vent, it seems more reasonable to suppose that they do not belong to it, but have arisen from a subsequent and an independent action.

We find scarcely an instance of a small crater occurring just in the centre of a large one, or taking the place of the cone. This is a curious circumstance. Whenever we have any central feature in a great crater that feature is a cone. The tendency of this fact is to prove that cones were produced by very weak efforts of this expiring force, for had there been any strength in the last paroxysm it is presumable that it would have blown out and left a crater. No very violent eruptions have therefore taken place from the vents that were connected with the great craters of the moon, nothing more powerful than could produce a cone of exudation or a cinder-heap. And with regard to cones, it is noteworthy that whether they be single or multiple, they never rise so high as the circular ramparts of their respective craters. This supports the inferred connexion between the crater origin and the cone origin, for supposing the two to have been independent, a supposition untenable in view of the universality of the central position of the cone, it is scarcely conceivable that the mountains should have always been located within ramparts higher than themselves. The less height argues less power in the upcasting agency, and the diminished force may well be considered as that which would almost of necessity precede the expiration of the eruption.

Occasionally a crater is met with that has a double rampart, and the concentricity suggests that there have been two eruptions from the same vent : one powerful, which formed the exterior circle, and a second rather less powerful which has formed the interior circle. It is not, however, evident that this duplication of the ring has always been due to a double eruption. In many cases there is duplication of only a portion : a terrace exhibits itself around a part of the circular range, sometimes upon the outside and sometimes upon the inside. These terraces are not likely to have been formed by any freak of the eruption, and we are led to ascribe them in general to landslip phenomena. When, in the course of a volcano's formation, the piling-up of material about the vent has continued till the lower portions have been unable to support the upper, or when from any cause, the material composing the pile has lost its cohesiveness, the natural consequence has been a breaking away of a portion of the structure and its precipitation down the inclined sides of the crater. Vast segments of many of the lunar mountain-rings appear to have been thus dislodged from their original sites and cast down the flanks to form crescent ranges of volcanic rocks either within or without the circle. Nearly every one of our plates contains craters exhibiting this feature in more or less extensive degree. Sometimes the separated portion has been very small in proportion to the circumference of the crater : Plato is an instance in which a comparatively small mass has been detached. In other cases very large segments have slid down and lie in segmental masses on the plateaux or form terraces around the rampart. Aristarchus, Treisnecker, and Copernicus exhibit this larger extent of dislocation.

It is possible that these landslips occurred long after the formation of the craters that have been subject to them. They are probably attributable to recent disintegration of the lunar rocks, and we have a powerful cause for this in the alternations of temperature to which the lunar crust is exposed. We shall have occasion to revert to this subject by-and-by; at present it must suffice to point out that the extremes of cold and heat, between which the lunar soil varies, are, with reasonable probability, assumed to be on the one hand the temperature of space (which is supposed to be about 200° below zero), and, on the other hand, a degree of heat equal to about twice that of boiling water. A range of at least 500° must work great changes in such heterogeneous materials as we may conjecture those of the lunar crust to be, by the alternate contractions and expansions which it must engender, and which must tend to enlarge existing fissures and create new ones, to grind contiguous surfaces and to dislodge unstable masses. This cause of change, it is to be remarked, is one which is still exerting itself.

In a few cases we have an entirely opposite interruption of the uniformity of a crater's contour. Instead of the breaking away of the ring in segments, we see the entire circuit marked with deep ruts that run down the flanks in a radial direction, giving us evidence of a downward *streaming* of semi-fluid matter, instead of a disruption of solid masses. We cannot doubt that these ruts have been formed by lava currents, and they indicate a condition of ejected material different from that which existed in the cases where the landslip character is found. In these last the ejecta appears to have been in the form of masses of solidified or rapidly solidifying matter, which remained where deposited for a time and then gave way from overloading or loss of cohesiveness, whereas the substances thrown out in the case of the rutted banks were probably mixed solid and fluid, the former remaining upon the flanks while the latter trickled away. Nothing so well represents, upon a small scale, this radial channelling as a heap of wetted sand left for a while for the water to

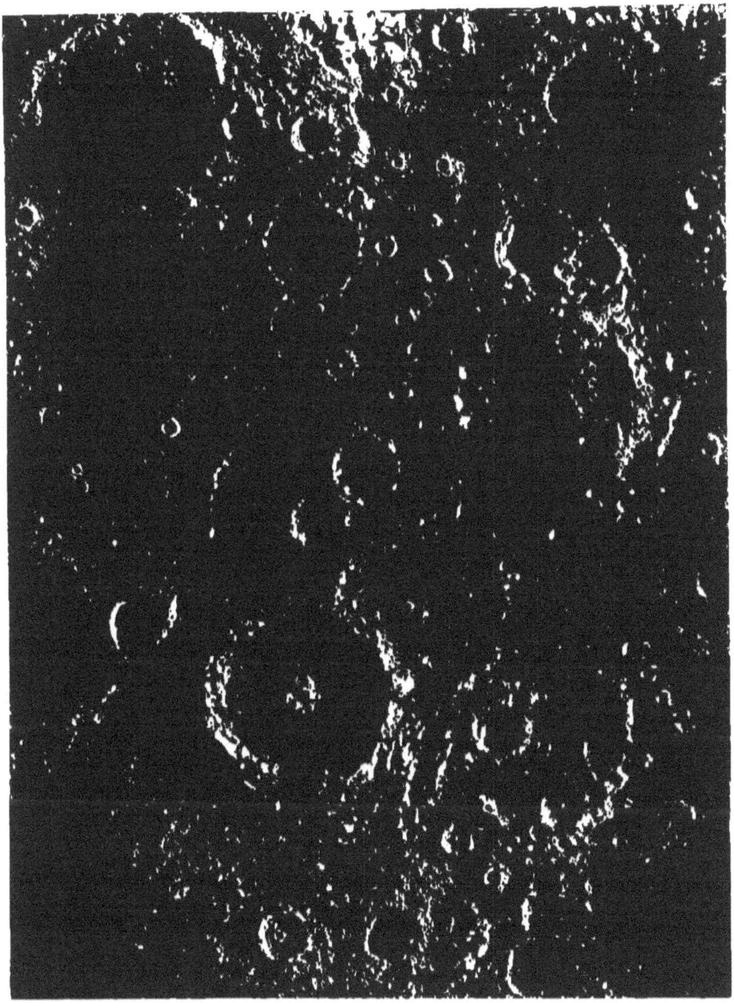

drain off from it. The solid grains in such a heap sustain its general mass-form, but the liquid in passing away cuts the surface into fissures running from the summit to the base, and forms it into a model of a volcanic mountain with every feature of peak, crag, and chasm reproduced. This similarity of effect leads us to suspect a parallelism of cause, and thus to the inference that the material which originally formed such a crater-mountain as Aristillus (which is a most prominent example of this rutted character, and appears in Plate IX., side by side with a crater that has its banks segmentally broken), must have been of the compound nature indicated ; and that an action analogous to that which ruts a damp sand-heap, rutted also the banks of the lunar crater.

Before passing from the subject of craters it behoves us to say a few words upon the curious manner in which these formations are complicated by intermingling and superposition. Yet, upon this point, we may be brief, for in the way of description our plates speak more forcibly than is possible by words. In particular we would refer to Plate XII., which represents the conspicuous group of craters of which the three largest members have been respectively named Theophilus, Cyrillus, and Catherina. But the area included in this plate is by no means an extraordinary one ; there are regions about Tycho wherein the craters so crowd and elbow each other that, in their intricate combinations, they almost defy accurate depiction. Our map and Plate XVI. will serve to give some idea of them. This intermingling of craters obviously shows that all the lunar volcanoes were not simultaneously produced, but that after one had been formed, an eruption occurred in its immediate neighbourhood and blew a portion of it away ; or it may have been that the same deep-seated vent at different times gave forth discharges of material the courses of which were more or less diverted on their way to the surface.

We have before alluded to the frequent occurrence of lines of craters upon the moon. In these lines the overlapping is frequently visible ; it is seen in Plate XII. before referred to, where the ring mountains are linked into a chain slightly curved, and upon the map, Plate IV., the nearly central craters Ptolemy and Alphonsus, the latter of which overlaps the former, are seen to form part of a line of craters marking a connection

Q

of primary disturbance. An extensive crack suggests itself as a favourable cause for the production of this overlaying of craters, for it would serve as a sort of " line of fire " from various points at which eruptions would burst forth, sometimes weak or far apart, when the result would be lines of isolated craters, and sometimes near together, or powerful, when the consequence would be the intrusion of one upon the other, and the perfect production of the latest formed at the expense or to the detriment of those that had been formed previously. The linear grouping of volcanoes upon the earth long ago struck observant minds. The fable of the *Typhon* lying under Sicily and the Phlegreian fields and disturbing the earth by its writhings, is a mythological attempt to explain the particular case in that region.

The capricious manner in which these intrusions occur is very curious. Very commonly a small crater appears upon the very rampart of a greater one, and a more diminutive one still will appear upon the rampart of the parasite. Stoeffler presents us with one example of this character, Hipparchus with another, Maurolycus with a third, and these are but a few cases of many. Here and there we observe several craters ranged in a line with their rims in one direction all perfect, and the whole appearing like a row of coins that have fallen from a heap. There is an example near to Tycho which we reproduce in Plate XX. In this case one is led to conjecture that the ejective agency, after exerting itself in one spot, travelled onward and renewed itself for a time ; that it ceased after forming crater number two, and again journeyed forward in the same line, recommencing action some miles further, and again subsiding ; yet again pushing forward and repeating its outburst, till it produced the fourth crater, when its power became expended. In each of these successive eruptions the centre of discharge has been just outside the crater last formed ; and the close connexion of the members of the group, together with the fact of their nearly similar size, appears to indicate a community of origin. For it seems feasible that as a general rule the size of a crater may be taken as a measure of the depth of force that gave rise to the eruption producing it. This may not be true for particular cases, but it will hold where a great number are collectively considered ; for if we assume the existence of an average disturbing force, it is apparently clear that it will manifest itself in disturbing greater or less surface-areas in proportion as it acts from greater or less depths. Or, *mutatis mutandis*, if we assume an uniform depth for the source of action, the greater or less surface disturbance will be a measure of greater or less eruptive intensity.

Perhaps the most remarkable case of a vast number of craters, which, from their uniform dimensions, suggest the idea of community of source-power or source-depth, is that offered by the region surrounding Copernicus, which, as will be seen by our plate of that object, is a vast Phlegreian field of diminutive craters. So countless are the minute craters that a high magnifying power brings into view when atmospheric circumstances are favourable, and so closely are they crowded together, that the resulting appearance suggests the idea of froth, and we should be disposed to christen this the "frothy region" of the moon, did not a danger exist in the tendency to connect a name with a cause. The craters that are here so abundant are doubtless the remains of true volcanoes analogous to the parasitical cones that are to be found on several terrestrial mountains, and not such accidental formations as the *Hornitos* described by Humboldt as abounding in the neighbourhood of the Mexican volcano, Jurillo, but which the traveller did not consider to be true cones of eruption.* Although upon our plate, and in comparison with the great crater that is its chief feature, these countless hollows appear so small as at first sight to appear insignificant, we must remember that the minutest of them must be grand objects, each probably equal in dimensions to Vesuvius. For since, as we have shown in an early chapter, the smallest discernible telescopic object must subtend an angle to our eye of about a second, and since this angle extended to the moon represents a mile of its surface, it follows that these tiny specks of shadow that besprinkle our picture, are in the reality craters of a mile diameter. This comparison may help the conception of the stupendous magnitude of the moon's volcanic features; for it is a conception most difficult to realize. It is hard to bring the mind to grasp the fact that that hollow of Copernicus is fifty miles in diameter. We read of an army having encamped in the once peaceful crater of Vesuvius, and of

* "Cosmos," Bohn's Edition, Vol. V. p. 322.

one of the extinct volcanoes of the *Campi Phlegræi* being used as a hunting preserve by an Italian king. These facts give an idea of vastness to those who have not the good fortune to see the actual dimensions of a volcanic orifice themselves. But it is almost impossible to conjure up a vision of what that fifty-mile crater would look like upon the moon itself; and for want of a terrestrial object as a standard of comparison, our picture, and even the telescopic view of the moon itself, fails to render the imagination any help. We may try to realize the vastness by considering that one of our average English counties could be contained within its ramparts, or by conceiving a mountainous amphitheatre whose opposite sides are as far apart as the cathedrals of London and Canterbury, but even these comparisons leave us unimpressed with the true majesty which the object would present to a spectator upon the surface of our satellite.

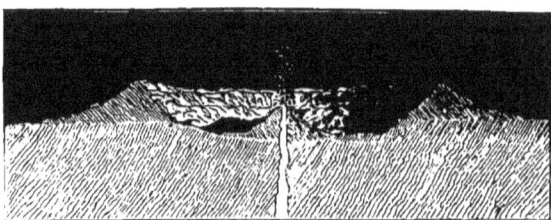

THE FORMATION OF THE CENTRAL CONE. FINAL ACTION OF A LUNAR VOLCANO.

CHAPTER IX.

ON THE GREAT RING-FORMATIONS NOT MANIFESTLY VOLCANIC.

IN our previous chapter we have given a reason for regarding as true volcanic craters all those circular formations, of whatever size, that exhibit that distinctive feature *the central cone*. Between the smallest crater with a cone that we can detect under the best telescopic conditions, namely, the companion to Hell, 1¾ mile diameter, and the great one called Petavius, 78 miles in diameter, we find no break in the continuity of the crater-cum-cone system that would justify us in saying that on the one side the volcanic or eruptive cause ceased, and on the other side some other causative action began. But there are numerous circular formations that surpass the magnitude of Petavius and its peers, but that have no circular cone, and are, therefore, not so manifestly volcanic as those which possess this feature. Our map will show many striking examples of this class at a glance. We may in particular refer *inter alia* to Ptolemy near the centre of the moon, to Grimaldi (No. 125), Shickard (No. 28), Schiller (No. 24), and Clavius (No. 13), all of which exceed 100 miles in diameter. Even the great *Mare Crisium*, nearly 300 miles in diameter, appears to be a formation not distinct from those which we have just named. These present little of the generic crater character in their appearance; and they have been distinguished therefrom by the name of *Walled* or *Ramparted Plains*. Their actual origin is beyond our explanation, and in attempting to account for them we must perforce allow considerable freedom to conjecture. They certainly, as Hooke suggested, present a "broken bubble"-like aspect; but one cannot reasonably imagine the existence of any form of mineral matter that would sustain itself in bubble form over areas of many hundreds of square miles. And if it were reasonable to suppose the great rings to be the foundations of such vast volcanic domes, we must conclude these to have broken when they could no longer sustain themselves, and in that case the surface beneath should be strewed with *débris*, of which, however, we can find no trace. Moreover, we might fairly expect that some of the smaller domes would have remained standing : we need hardly say that nothing of the kind exists.

The true circularity of these objects appears at first view a remarkable feature. But it ceases to be so if we suppose them to have been produced by some very concentrated sublunar force of an upheaving nature, and if

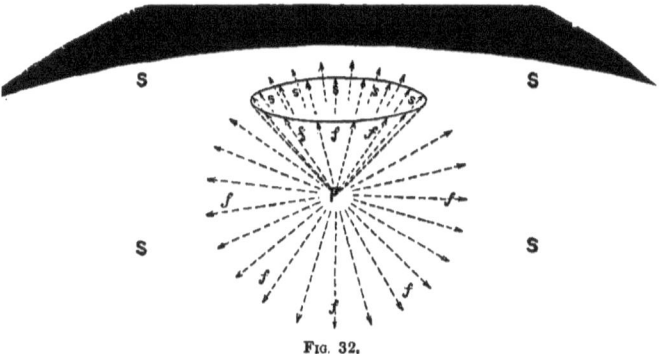

FIG. 32.

only we admit the homogeneity of the moon's crust. For if the crust be homogeneous, then *any* upheaving force, deeply seated beneath it, will exert itself *with equal effects at equal distances from the source :* the lines of equal effect will obviously be radii of a sphere with the source of the disturbance for its centre, and they will meet a surface over the source in a circle. This will be evident from Fig. 32, in which a force is supposed to act at F below the surface s s s s. The matter composing s s being homogeneous, the action of F will be equal at equal distances in all directions. The lines of equal force, F f, F f, will be of equal length, and they will form, so to speak, radii of a sphere of force. This sphere is cut by the plane at $s\ s\ s\ s$, and as the intersection necessarily takes place everywhere at the extremity of these radii, the figure of intersection is demonstrably a circle (shown in perspective as an ellipse in the figure). Thus we see that an intense but extremely confined explosion, for instance, beneath the moon's crust must disturb a *circular* area of its surface, if the intervening material be homogeneous. If this be not homogeneous there would be, where it offered *less* than the average resistance to the disturbance, an outward distortion of the circle ; and an opposite interruption to circularity if it offers *more* than the average resistance. This assumed homogeneity may possibly be the explanation of the general circularity of the lunar surface features, small and great.

We confess to a difficulty in accounting for such a very local generation of a deep-seated force ; and, granting its occurrence, we are unprepared with a satisfactory theory to explain the resultant effect of such a force in producing a raised ring at the limit of the circular disturbance. We may, indeed, suppose that a vast circular cake or conical frustra would be

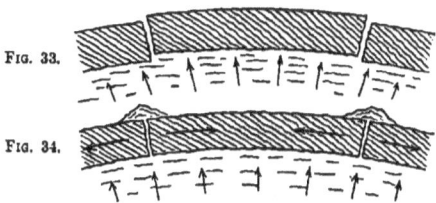

Fig. 33.

Fig. 34.

temporarily upraised as in Fig. 33, and that upon its subsidence a certain extrusion of subsurface matter would occur around the line or zone of rupture as in Fig. 34. This supposition, however, implies such a peculiarly cohesive condition of the matter of the uplifted cake, that it is doubtful whether it can be considered tenable. We should expect any ordinary form of rocky matter subjected to such an upheaval to be fractured and distorted, especially when the original disturbing force is greater in the centre than at the edge, as, according to the above hypothesis, it would be ; and in subsiding, the rocky plateau would thus retain some traces of its disturbance ; but in the circular areas upon the moon there is nothing to indicate that they have been subjected to such dislocations.

Mr. Scrope in his work on volcanoes has given a hypothetical section of a portion of the earth's crust, which presents a bulging or tumescent surface in some measure resembling the effect which such a cause as we have been considering would produce. We give a slightly modified version of his sketch in Fig. 35, showing what would be the probable

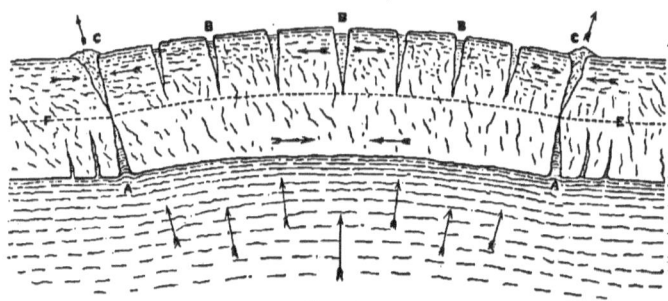

FIG. 35.

A A. Fissures gaping downwards and injected by intumescent lava beneath. D D D. Fissures gaping upwards and allowing wedges of rock to drop below the level of the intervening masses. C C. Wedges forced upwards by horizontal compression. E F. Neutral plane or pivot axis, above and below which the directions of the tearing strain and horizontal compression are severally indicated by the smaller arrows; the larger arrows beneath represent the direction of the primary expansive force.

phenomena attending such an upheaval as regards the behaviour of the disturbed portion of the crust, and also that of the lava or semifluid matter beneath: and, as will be seen by the sketch, a possible phase of the phenomena is the production of an elevated ridge or rampart at the points of disruption *c c*; and where there is a ring of disruption, as by our hypothesis there would be, the ridge or rampart *c c* would be a circle. In this drawing we see the cracking and distortion to which the elevated area would be subjected, but of which, as previously remarked, the circular areas of the moon present no trace of residual appearance.

Those who have offered other explanations of these vast ring-formed mountain ranges, have been no more happy in their conjectures. M. Rozet, who communicated a paper on selenology to the French

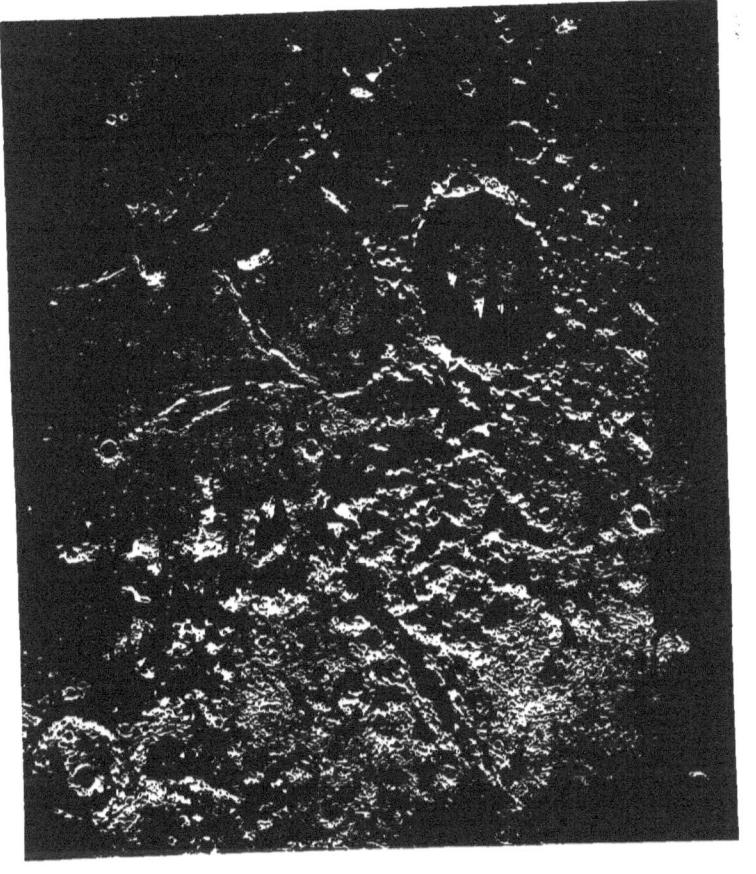

Academy in 1846, put forth the following theory. He argued that during the formation of the solid scoriaceous pelicules of the moon, circular or tourbillonic movements were set up; and these, by throwing the scoria from the centre to the circumference, caused an accumulation thereof at the limit of the circulation. He considered that this phenomenon continued during the whole process of solidification, but that the amplitude of the whirlpool diminished with the decreasing fluidity of the surface material. Further, he suggested that when many vortices were formed, and the distances of their centres, taken two and two, were less than the sums of their radii, there resulted close spaces terminated by arcs of circles; and when for any two centres the distance was greater than the sum of their radii of action, two separate and complete rings were formed. We have only to remark on this, that we are at a loss to account for the origination of such vorticose movements, and M. Rozet is silent on the point. If the great circles are to be referred to an original sea of molten matter, it appears to us more feasible to consider that wherever we see one of them there has been, at the centre of the ring, a great outflow of lava that has flooded the surrounding surface. Then, if from any cause, and it is not difficult to assign one, the outflow became intermittent, or spasmodic, or subject to sudden impulses, concentric waves would be propagated over the pool and would throw up the scoria or the solidifying lava in a circular bank at the limit of the fluid area.

This hypothesis does not differ greatly from the *ebullition* theory proposed by Professor Dana, the American geologist, to explain these formations. He considered that the lunar ring-mountains were formed by an action analogous to that which is exemplified on the earth in the crater of Kilauea, in the Hawaiian islands. This crater is a large open pit exceeding three miles in its longer diameter, and nearly a thousand feet deep. It has clear bluff walls round a greater part of its circuit, with an inner ledge or plain at their base, raised 340 feet above the bottom. This bottom is a plain of solid lavas, entirely open to-day, which may be traversed with safety (we are quoting Professor Dana's own statement written in 1846, and therefore not correctly applying to the present time): over it there are pools of boiling lava in active ebullition, and one is more than a thousand feet in diameter. There are also cones at times, from a few yards to two or three thousand feet in diameter, and varying greatly in angle of inclination. The largest of these cones have a circular pit or crater at the summit. The great pit itself is oblong, owing to its situation on a fissure, but the lakes upon its bottom are round, and in them, says Professor Dana, " the circular or slightly elliptical form of the moon's craters is exemplified to perfection."

Now Dana refers this great pit crater and its contained lava-lakes to "the fact that the action at Kilauea is simply *boiling*, owing to the extreme fluidity of the lavas. The gases or vapours which produce the state of active ebullition escape freely in small bubbles, with little commotion, like jets over boiling water; while at Vesuvius and other like cones they collect in immense bubbles before they accumulate force enough to make their way through; and consequently the lavas in the latter case are ejected with so much violence that they rise to a height often of many thousand feet and fall around in cinders. This action builds up the pointed mountain, while the simple boiling of Kilauea makes no cinders and no cinder cones."

Professor Dana continues, "If the fluidity of lavas, then, is sufficient for this active ebullition, we may have boiling going on over an area of an indefinite extent; for the size of a boiling lake can have no limits except such as may arise from a deficiency of heat. The size of the lunar craters is therefore no mystery. Neither is their circular form difficult of explanation; for a boiling pool necessarily, by its own action, extends itself circularly around its centre. The combination of many circles, and the large sea-like areas are as readily understood." *

In justice to Professor Dana it should be stated that he included in this theory of formation all lunar craters, even those of small size and possessing central cones; and he put forth his views in opposition to the eruptive theory which we have set forth, and which was briefly given to the world more than twenty-five years ago. As regards the smallest craters with cones, we believe few geologists will refuse their compliance with the supposition that they were formed as our crater-bearing volcanoes were formed: and we have pointed out the logical impossibility of

* *American Journal of Science, Second Series, Vol. II.*

assigning any limit of size beyond which the eruptive action could not be said to hold good, so long as the central cone is present. But when we come to ring-mountains having no cones, and of such enormous size that we are compelled to hesitate in ascribing them to ejective action, we are obliged to face the possibility of some other causation. And, failing an explanation of our own that satisfied us, we have alluded to the few hypotheses proffered by others, and of these Professor Dana's appears the most rational, since it is based upon a parallel found on the earth. In citing it, however, we do necessarily not endorse it.

CHAPTER X.

PEAKS AND MOUNTAIN RANGES.

THE lunar features next in order of conspicuity are the mountain ranges, peaks, and hill-chains, a class of eminences more in common with terrestrial formations than the craters and circular structures that have engaged our notice in the preceding chapters.

In turning our attention to these features, we are at the outset struck with the paucity on the lunar surface of extensive mountain systems as compared with its richness in respect of crateral formations; and a field of speculation is opened by the recognition of the remarkable contrast which the moon thus presents to the earth, where mountain ranges are the rule, and craters like the lunar ones are decidedly exceptional. Another conspicuous but inexplicable fact is that the most important ranges upon the moon occur in the northern half of the visible hemisphere, where the craters are fewest and the comparatively featureless districts termed "seas" are found. The finest range is that named after our Apennines and which is included in our illustrative Plate, No. IX. It extends for about 450 miles and has been estimated to contain upwards of 3000 peaks, one of which—Mount Huyghens—attains the altitude of 18,000 feet. The Caucasus is another lunar range which appears like a diverted northward extension of the Apennines, and, although a far less imposing group than the last named, contains many lofty peaks, one of which approaches the altitude assigned to Mount Huyghens while several others range between 11,000 and 14,000 feet high. Another considerable range is the Alps, situated between the Caucasus and the crater Plato, and reproduced on Plate XIV. It contains some 700 peaked mountains and is remarkable for the immense valley, 80 miles long and about five

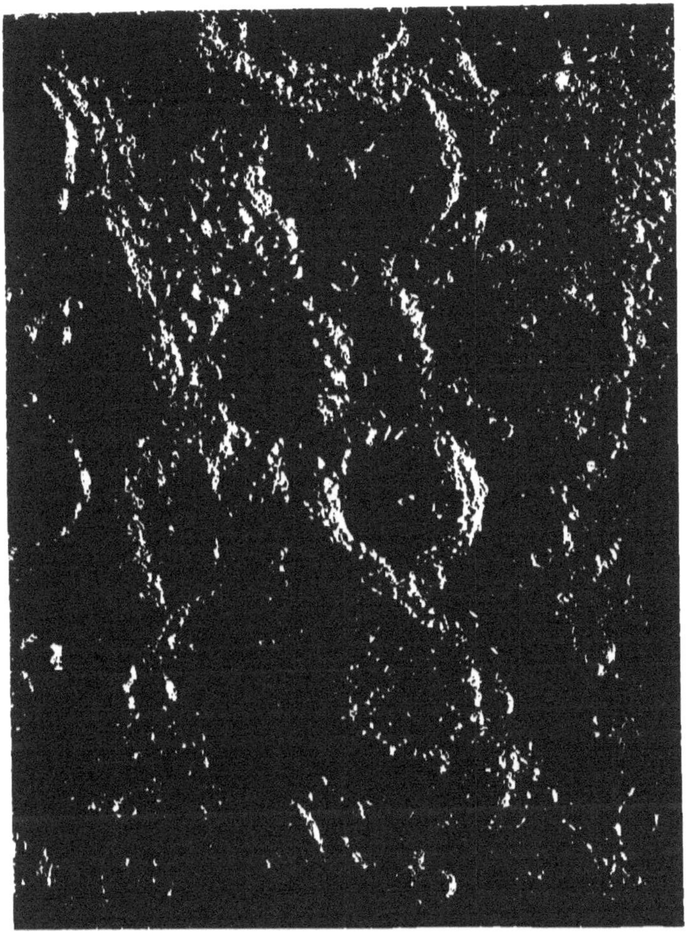

PLATE XV

MERCATOR & CAMPANUS.

SCALE

broad, that cuts it with seemingly artificial straightness; and that, were it not for the flatness of its bottom, might set one speculating upon the probability of some extraneous body having rushed by the moon at an enormous velocity, gouging the surface tangentially at this point and cutting a channel through the impeding mass of mountains. There are other mountain ranges of less magnitude than the foregoing; but those we have specified will suffice to illustrate our suggestions concerning this class of features.

We remark, too, that there is a prevailing tendency of the ranges just mentioned to present their loftiest constituents in abrupt terminal lines, facing nearly the same direction, the reverse of that towards which they are carried by the moon's rotation; and as they recede from the high terminal line, the mountains gradually fall off in height, so that in bulk the ranges present the "crag and tail" contour which individual hills upon the earth so frequently exhibit.

Isolated peaks are found in small numbers upon the moon; there are a few striking examples of them nevertheless, and these are chiefly situated in the mountainous region just alluded to. Several are seen to the east (right hand) of the Alpine range depicted on Plate XIV. The best known of these is Pico, which rises abruptly from a generally smooth plain to a height of 7000 feet. It may be recognized as the lower of the two long shadowing spots located almost centrally above the crater Plato in the illustration just mentioned. Above it, at an actual distance of 40 miles, there is another peak (unnamed) about 4000 feet high; and away to the west, beyond the small crater joined by a hill-ridge to Plato, is a third pyramidal mountain nearly as high as Pico.

It seems natural to regard the great mountain chains as agglomerations of those peaks of which we have isolated examples in Pico and its compeers, and thus to consider that the formation of a mountain chain has been a multiplication of the process that formed the single pyramid-shaped eminences. At first thought it might appear that the great mountain ranges were produced by bodily upthrustings of the crust of the moon by some subsurface convulsions. But such an explanation could hardly hold in relation to the isolated peaks, for it is difficult, if not impossible, to conceive that these abrupt mountains, almost resembling a sugarloaf in steepness, could have been protruded *en masse* through a smooth region of the crust. On the contrary it is quite consistent with probability to suppose that they were built up by a slow process somewhat analogous to that to which we have ascribed the piling of the central cones of the greater craters. We believe they may be regarded as true mountains of exudation, produced by the comparatively gentle oozing of lava from a small orifice and its solidification around it; the vent however remaining open and the summit or discharging orifice continually rising with the growth of the mountain, as indicated in the annexed cut, Fig. 36. This process is well exemplified in the case of a

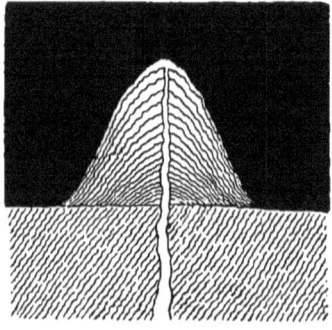

FIG. 36.

water fountain playing during a severe frost; the water as it falls around the lips of the orifice freezes into a hillock of ice, through the centre of which, however, a vent for the fluid is preserved. As the water trickles over the mound it is piled higher and higher by accumulating layers of ice, till at length a massive cone is formed whose height will be determined by the force or "head" of the water. Substitute lava for water, and we have at once a formative process which may very fairly be considered as that which has given rise to the isolated mountains of the moon.

There are upon the earth mountainous forms resembling the isolated peaks of the moon, and which have been explained by a similar theory to the above. We reproduce a figure of one observed by Dana at Hawaii

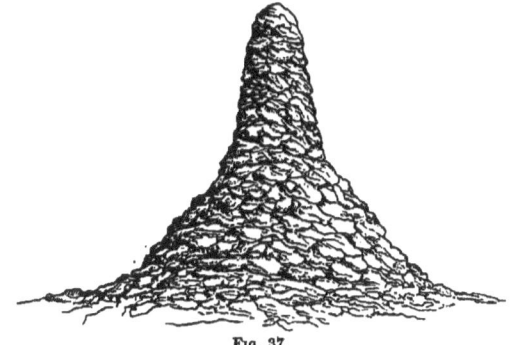

Fig. 37.

(Fig. 37), and a sketch of another observed on the summit of the Volcano of Bourbon (Fig. 38); we also reproduce (Fig. 39) an ideal section of the latter, given by Mr. Scrope, and showing the successive

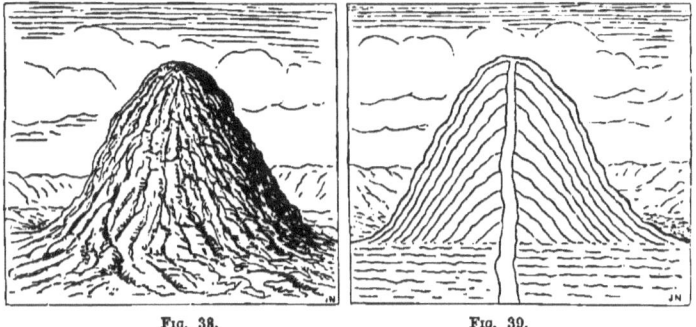

Fig. 38.   Fig. 39.

layers of lava which would be disposed by just such an action as that manifested in the case of the freezing fountain; and we quote that author's words in reference to this explanation of the formation of

Etna and other volcanic mountains. "On examining," says Mr. Scrope,* "the structure of the mountain (Etna) we find its entire mass, so far as it is exposed to view by denudation or other causes (and one enormous cavity, the Val de Bove penetrates deeply into its very heart), to be composed of beds of lava-rock alternating more or less irregularly with layers of scoriæ, lapillo and ashes, almost precisely identical in mineral character, as well as in general disposition, with those erupted by the volcano at known dates within the historical period. Hence we are fully justified in believing the whole mountain to have been built up in the course of ages in a similar manner by repeated intermittent eruptions. And the argument applies by the rules of analogy to all other volcanic mountains, though the history of their recent eruptions may not be so well recorded, provided that their structure corresponds with, and can be fairly explained by, this mode of production. It is also further applicable, under the same reservation, to all mountains composed entirely, or for the most part, of volcanic rocks, even though they may not have been in eruption within our time."

To these illustrations furnished from Scrope's work we add another, copied from a photograph by Professor Piazzi Smyth, of a "blowing cone" at the base of Teneriffe (Fig. 40), which is but one of many that are to be found on that mountain and which has been formed by a process similar to that we have been considering, but acting upon a comparatively small scale. Professor Smyth describes this cone as about 70 feet high and of parabolic figure, composed of hard lava and with an upper aperture still yawning, "whence the burning breath of fires beneath once issued in fury and with destruction."

Reverting now to the moon, we remark that, if the foregoing explanation of the isolated lunar peaks be tenable, it should hold equally for the groups of them which we see in the lunar Apennines, Alps, Caucasus, and other ranges of like character. There occur in some places intermediate groups which link the one to the other. Just above the crater Archimedes, on Plate IX., for instance, we see several single peaks and small clumps of them leading by successive multiple-peak examples to

* "Volcanoes," page 155.

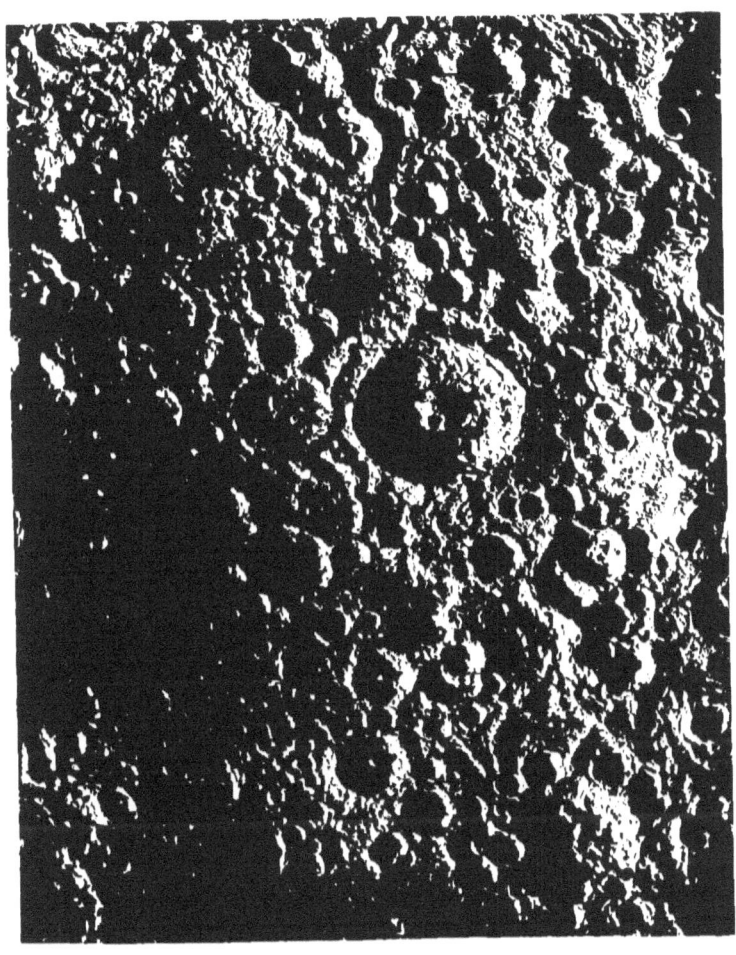

what may be called chains of mountains like many that are included in the contiguous Apennine system. And, in view of this connexion between the single peaks and the mountain ranges formed of aggre-

FIG. 40.
SMALL VOLCANIC MOUNTAIN AT THE END OF A STREET AT TENERIFFE.

gations of such peaks, it seems to us reasonable to conclude that the latter were formed by the comparatively slow escape of lava through multitudinous openings in a weak part of the moon's crust, rather than to suppose that the crust itself has been bodily upheaved and retained in its disturbed position. The high peaks that many mountains in such a chain exhibit accord better with the former than the latter explanation; for it is difficult to imagine how such lofty eminences could be erected by an upheaval, and we must remember that the moon has none of the denuding elements which are at work upon the earth, weather-wearing its mountain forms into sharpness and steepness.°

* In reference to such prominences on the lunar surface as cast steeple-like shadows, it is well to remark that we must not in all cases infer, from the acute spire-like form of the shadow, that the object which casts the shadow is of a similar sharp or spire-like form, which the first

And we have ground for believing the mountain-forming process on the moon to have been a comparatively gentle one, in the fact that the mountain systems appear in regions otherwise little disturbed, and where craters, which have all the appearances of violent origin, are few and far between. Evidently the mountain and crater-forming processes, although both due to extrusive action, were in some measure different, and it is reasonable to suppose that the difference was in degree of intensity; so that while a violent ejection of volcanic material would give rise to a crater, a more gradual discharge would pile up a mountain. In this view craters are evidences of *eruptive*, and mountains of comparatively gentle *exudative* action.

We can hardly speculate with any degree of safety upon the cause of this varying intensity of volcanic discharge. We may ascribe it to variation of *depth* of the initial disturbing force, or to suddenness of its action; or it may be that different degrees of fluidity of the lava have had modifying effects; or on the other hand different qualities of the crust-material; or yet again differences of period—the quieter extrusions having occurred at a time when the volcanic forces were dying down. There is an alliance between lunar craters and mountains that goes far to show that there has been no radical difference in their origins. For instance, as we have previously pointed out, craters in some cases run in

impression would naturally lead us to suppose. A comparatively blunt or rounded eminence will project a long and pointed shadow when the rays of light fall on the object at a low angle, and especially so when the shadow is projected on a convex surface. We illustrate this with a copy of an actual photograph of the shadow cast by half a pea, Fig. 41.

FIG. 41.

linear groups, as if in those cases they had been formed along a line of disruption or of least resistance of the crust; and the mountain chains have a corresponding linear arrangement. Then we see craters and mountain chains disposed in what seem obviously the same arcs of disturbance. Thus Copernicus (No. 147), Erastothenes (No. 168), and the Apennines appear to belong to one continuous line of eruption; and it requires no great stretch of imagination to suppose that the Caucasus, Eudoxus (No. 208), and Aristotle (No. 209) form a continuation of the same line. Then around the Mare Serenetatis we see mountainous ridges and craters alternating one with the other as though the exuding action there, normally sufficient to produce the ridges, had at some points become forcible enough to produce a crater. Again, upon the very mountain ranges themselves, as for instance among the Apennines, we find small craters occurring. We see, too, that the great craters are in many cases surrounded by radiating systems of ridges which almost assume mountainous proportions, and which are doubtless exuded matter from "starred" cracks, the centres of which are occupied by the craters. The same kind of ridges here and there occur apart from craters (see for instance Plate XVIII., below Aristarchus and Herodotus) and sometimes they occur in the neighbourhood of extensive cracks, to which they also seem allied. We must indeed regard a linear crack as the origin either of a ridge (if the exudation is slight) or of a mountain chain (if the exudation is more copious) or a string of craters (if the extrusion rises to eruptive violence.) But the subject of cracks is important enough to be treated in a separate chapter.

We alluded in Chap. III. to the phenomena of wrinkling or puckering as productive of certain mountainous formations; and we pointed out the striking similarity in character of configuration between a shrivelled skin and a terrestrial mountain region. We do not perceive upon the moon such a decided coincidence of appearances extending over any considerable portion of her surface; but there are numerous limited areas where we behold mountainous ridges which partake strongly of the wrinkle character; and in some cases it is difficult to decide whether the puckering agency or the exudative agency just discussed has produced the ridges. The district bordering upon Aristarchus and Herodotus, above referred to, is of this doubtful character; and a similar district is that contiguous to Triesnecker (Plate XI.). There are, however, abundant examples of less prominent lines of elevation, which may, with more probability, be ascribed to a veritable wrinkling or puckering action; they are found over nearly the whole lunar surface, some of them standing out in considerable relief, and some merely showing gentle lines of elevation, or giving the surface an undulating appearance. A close examination of our picture-map (Plate IV.) will reveal very numerous examples, especially in the south-east (right-hand-upper) quadrant. Some of these lines of tumescence are so slightly prominent that we may suppose them to have been caused by the action indicated by Fig. 6 (p. 28), while others, from their greater boldness, appear to indicate a formative action analogous to that represented by Fig. 9 (p. 29).

IDEAL SKETCH OF PICO AS IT WOULD PROBABLY APPEAR IF SEEN BY A SPECTATOR LOCATED ON THE MOON.

PLATE XVII

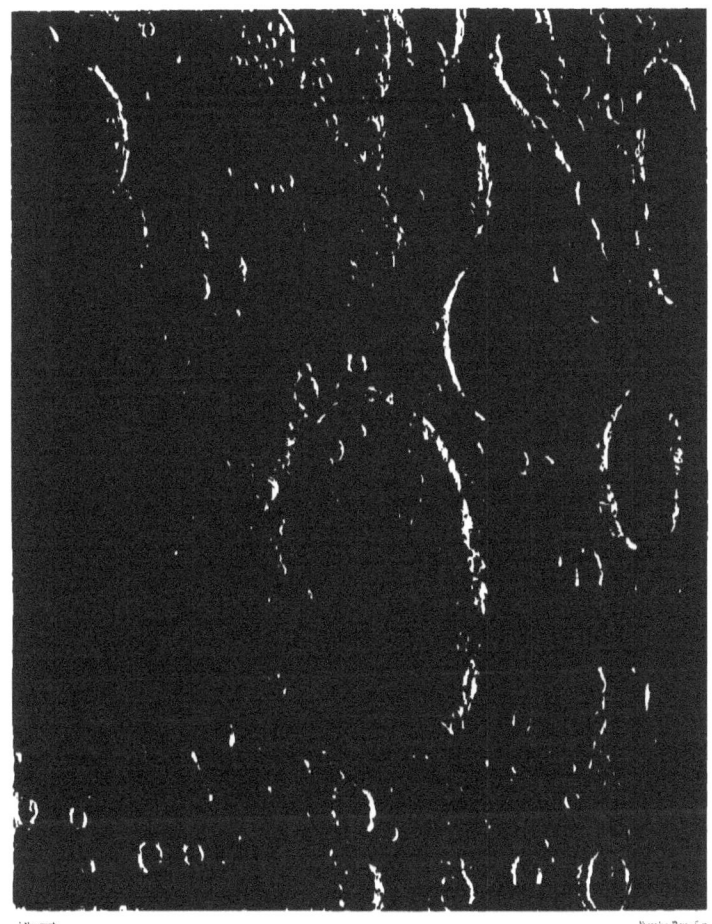

WARGENTIN.

SCALE.

CHAPTER XI.

CRACKS AND RADIATING STREAKS.

WE have hitherto confined our attention to those reactions of the moon's molten interior upon its exterior which have been accompanied by considerable extrusions of sub-surface material in its molten or semi-solid condition. We now pass to the consideration of some phenomena resulting in part from that reaction and in part from other effects of cooling, which have been accompanied by comparatively little ejection or upflow of molten matter, and in some cases by none at all. Of such the most conspicuous examples are those bright streaks that are seen, under certain conditions of illumination, to radiate in various directions from single craters, and some of the individual radial branches of which extend from four to seven hundred miles in a great arc on the moon's surface.

There are several prominent examples of these bright streak systems upon the visible hemisphere of the moon; the focal craters of the most conspicuous are Tycho, Copernicus, Kepler, Aristarchus, Menelaus, and Proclus. Generally these focal craters have ramparts and interiors distinguished by the same peculiar bright or highly reflective material which shows itself with such remarkable brilliance, especially at full moon: under other conditions of illumination they are not so strikingly visible. At or nearly full moon the streaks are seen to traverse over plains, mountains, craters, and all asperities; holding their way totally disregardful of every object that happens to lay in their course.

The most remarkable bright streak system is that diverging from the great crater Tycho. The streaks that can be easily individualized in this group number more than one hundred, while the courses of some of them may be traced through upwards of six hundred miles from their centre of divergence. Those around Copernicus, although less remarkable in regard to their extent than those diverging from Tycho, are nevertheless in many respects well deserving of careful examination : they are so numerous as utterly to defy attempts to count them, while their intricate reticulation renders any endeavour to delineate their arrangement equally hopeless.

The fact that these bright streaks are invariably found diverging from a crater, impressively indicates a close relationship or community of origin between the two phenomena : they are obviously the result of one and the same causative action. It is no less clear that the actuating cause or prime agency must have been very deep-seated and of enormous disruptive power to have operated over such vast areas as those through which many of the streaks extend. With a view to illustrate experimentally what we conceive to have been the nature of this actuating cause, we have taken a glass globe and, having filled it with water and hermetically sealed it, have plunged it into a warm bath : the enclosed water, expanding at a greater rate than the glass, exerts a disruptive force on the interior surface of the latter, the consequence being that at the point of least resistance, the globe is rent by a vast number of cracks diverging in every direction from the focus of disruption. The result is such a strikingly similar counterpart of the diverging bright streak systems which we see proceeding from Tycho and the other lunar craters before referred to, that it is impossible to resist the conclusion that the disruptive action which originated them operated in the same manner as in the case of our experimental illustration ; the disruptive force in the case of the moon being that to which we have frequently referred as due to the expansion which precedes the solidification of molten substances of volcanic character.

On Plate XIX. we present a photograph from one of many glass globes which we have cracked in the manner described : a careful comparison between the arrangement of divergent cracks displayed by this photograph with the streaks seen spreading from Tycho upon the contiguous lunar photograph will, we trust, justify us in what we have stated as to the similarity of the causes which have produced such identical results.

The accompanying figures will further illustrate our views upon the causative origin of the bright streaks. The primary action rent the solid crust of the moon and produced a system of radiating fissures (Fig. 42): these immediately afforded egress for the molten matter beneath to make its appearance on the surface simultaneously along the entire course of every crack, and irrespective of all surface inequalities

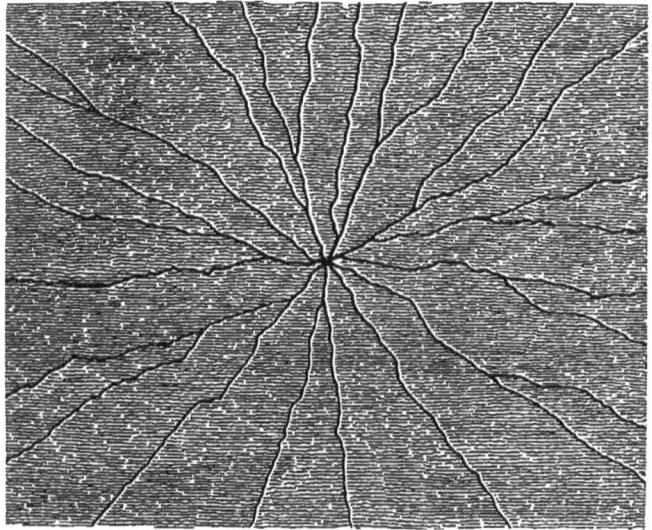

FIG. 42.
ILLUSTRATIVE OF THE RADIATING CRACKS WHICH PRECEDE THE FORMATION OF THE BRIGHT STREAKS.

or irregularities whatever (Fig. 43). We conceive that the upflowing matter spread in both directions sideways, and in this manner produced streaks of very much greater width than the cracks or fissures up through which it made its way to the surface.

In further elucidation of this part of our subject we may refer to a familiar but as we conceive cogent illustration of an analogous action in the behaviour of water beneath the ice of a frozen pond, which, on being fractured by some concentrated pressure, or by a blow, is well known to "star" into radiating or diverging cracks, up through which the water immediately issues, making its appearance on the surface of the ice simultaneously along the entire course of every crack, and on reaching

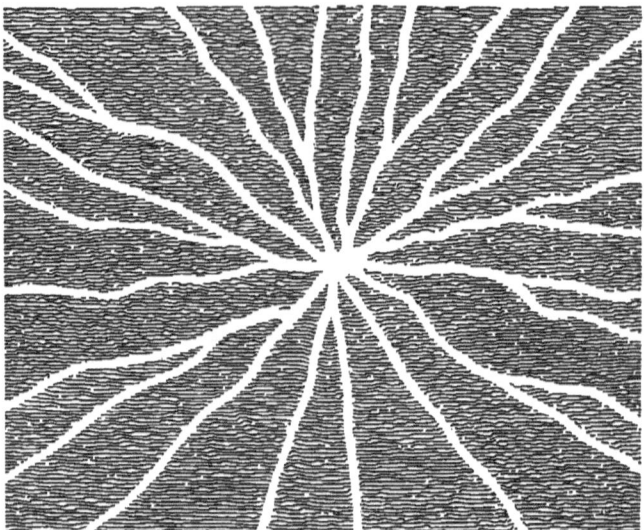

Fig. 43.
ILLUSTRATIVE OF THE RADIATING BRIGHT STREAKS.

the surface, spreading on both sides to a width much exceeding that of the crack itself.

If this familiar illustration be duly considered, we doubt not it will be found to throw considerable light on the nature of those actions which have resulted in the bright streaks on the moon's surface. Some have attempted to explain the cause of these bright streaks by assigning them to streams of lava, issuing from the crater at the centre of their divergence and flowing over the surface, but we consider such an explanation totally untenable, as any idea of lava, be it ever so fluid at its first

PLATE XVIII

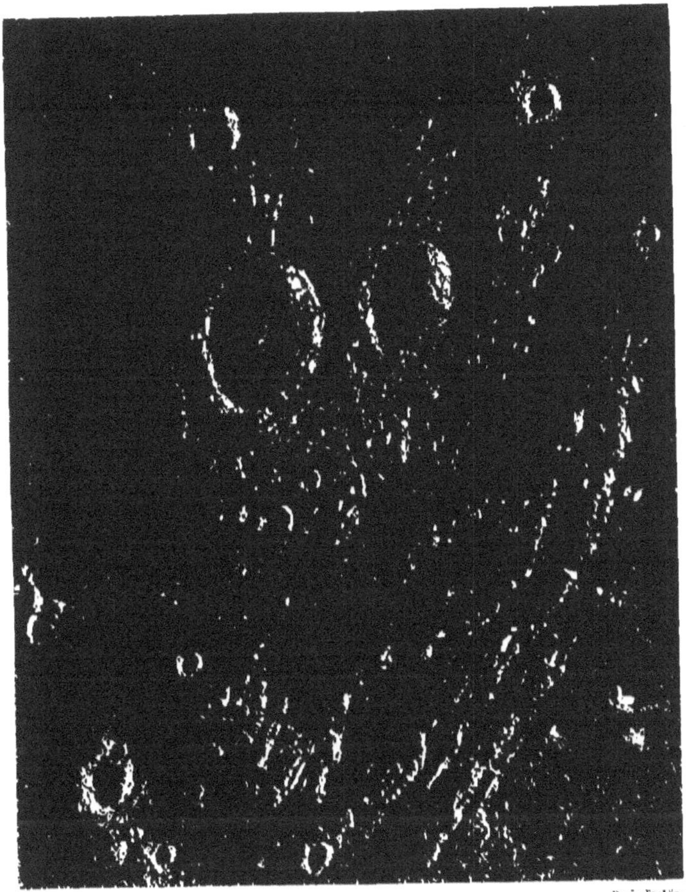

J Nasmyth

Brooks Day & Son

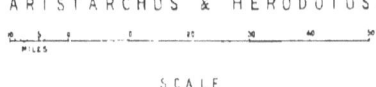

ARISTARCHUS & HERODOTUS.

SCALE

issue from its source, flowing in streams of nearly equal width, through courses several hundred miles long, up hills, over mountains, and across plains, appears to us beyond all rational probability.

It may be objected to our explanation of the formation of these bright streaks, that so far as our means of observation avail us, we fail to detect any shadows from them or from such marginal edges as might be expected to result from a side-way spreading outflow of lava from the cracks which afforded it exit in the manner described. Were the edges of these streaks terminated by cliff-like or craggy margins of such height as 30 or 40 feet, we might just be able at low angles of illumination and under the most favourable circumstances of vision, to detect some slight appearance of shadows; but so far as we are aware, no such shadows have been observed. We are led to suppose that the impossibility of detecting them is due not to their absence but to the height of the margins being so moderate as not to cast any cognizable shadow, inasmuch as an abrupt craggy margin of 10 or 15 feet high would, under even the most favourable circumstances, fail to render such visible to us. Reference to our ideal section of one of these bright streaks (Fig. 45), will show how thin their edges may be in relation to their spreading width.

The absence of cognizable shadows from the bright streaks has led some observers to conclude that they have no elevation above the surface over which they traverse, and it has therefore been suggested that their existence is due to possible vapours which may have issued through the cracks, and condensed in some sublimated or pulverulent form along their courses, the condensed vapours in question forming a surface of high reflective properties. That metallic or mineral substances of some kinds do deposit on condensation very white powders, or sublimates, we are quite ready to admit, and such explanation of the high luminosity of the bright streaks, and of the craters situated at the foci or centres of their divergence is by no means improbable, so far as concerns their mere brightness. But as we invariably find a crater occupying the centre of divergence, and such craters are possessed of all the characteristic features and details which establish their true volcanic nature as the results of energetic extrusions of lava and scoria, we cannot resist the conclusion that the material of the crater, and that of the bright streaks diverging

T

from it, are not only of a common origin, but are so far identical that the only difference in the structure of the one as compared with the other is due to the more copious egress of the extruded or erupted matter in the case of the crater, while the restricted outflow or ejection of the matter up through the cracks would cause its dispersion to be so comparatively gentle as to flood the sides of the cracks and spread in a thin sheet more or less sideways simultaneously along their courses. There are indeed evidences in the wider of the bright streaks of their being the result of the outflow of lava through *systems of cracks* running parallel to each other, the confluence of the lava issuing from which would naturally yield the appearance of one streak of great width. Some of those diverging from Tycho are of this class; many other examples might be cited, among which we may name the wide streaks proceeding from the crater Menelaus and also those from Proclus. Some of these occupy widths upwards of 25 miles—amply sufficient to admit of many concurrent cracks with confluent lava outflows.

We are disposed to consider as related to the fore-mentioned radiating streaks, the numerous, we may say the multitudinous, long and narrow chasms that have been sometimes called "canals" or "rills," but which are so obviously *cracks* or chasms, that it is desirable that this name should be applied to them rather than one which may mislead by implying an aqueous theory of formation. These cracks, singly and in groups, are found in great numbers in many parts of the moon's surface. As a few of the more conspicuous examples which our plates exhibit we may refer to the remarkable group west of Treisnecker (Plate XI.), the principal members of which converge to or cross at a small crater, and thus point to a continuity of causation therewith analogous to the evident relation between the bright streaks and their focal craters. Less remarkable, but no less interesting, are those individual examples that appear in the region north of (below) the Apennines (Plate IX.), and some of which by their parallelism of direction with the mountain-chain appear to point to a causative relation also. There is one long specimen, and several shorter in the immediate neighbourhood of Mercator and Campanus (Plate XV.); and another curious system of them, presenting suggestive contortions, occurs in connection with the mountains Aristarchus and

Herodotus (Plate XVIII.). Others, again, appear to be identified with the radial excrescences about Copernicus (Plate VIII.). Capuanus, Agrippa, and Gassendi, among other craters, have more or less notable cracks in their vicinities.

Some of these chasms are conspicuous enough to be seen with moderate telescopic means, and from this maximum degree of visibility there are all grades downwards to those that require the highest optical powers and the best circumstances for their detection. The earlier selenographers detected but a few of them. Schroeter noted only 11; Lohrman recorded 75 more; Beer and Maedler added 55 to the list, while Schmidt of Athens raised the known number to 425, of which he has published a descriptive catalogue. We take it that this increase of successive discoveries has been due to the progressive perfection of telescopes, or, perhaps, to increased education, so to speak, of the eye, since Schmidt's telescope is a much smaller instrument than that used by Beer and Maedler, and is regarded by its owner as an inferior one for its size. We doubt not that there are hundreds more of these cracks which more perfect instruments and still sharper eyes will bring to knowledge in the future.

While these chasms have all lengths from 150 miles (which is about the extent of those near Treisnecker) down to a few miles, they appear to have a less variable breadth, since we do not find many that at their maximum openings exceed two miles across; about a mile or less is their usual width throughout the greater part of their length, and generally they taper off to invisibility at their extremities, where they do not encounter and terminate at a crater or other asperity, which is, however, sometimes the case. Of their depth we can form no precise estimate, though from the sharpness of their edges we may conclude that their sides approach perpendicularity, and, therefore, that their depth is very great; we have elsewhere suggested ten miles as a possible profundity. In a few cases, and under very favourable circumstances, we have observed their generally black interiors to be interrupted here and there with bright spots suggestive of fragments from the sides of the cracks having fallen into the opening.

In seeking an explanation of these cracks, two possible causes suggest themselves. One is the expansion of subsurface matter, already suggested as explanatory of the bright streaks; the other, a contraction of the crust by cooling. We doubt not that both causes have been at work, one perhaps enhancing the other. Where, as in the cases we have pointed out, there are cracks which are so connected with craters as to imply relationship, we may conclude that an upheaving or expansive force in the sublunar molten matter has given rise to the cracks, and that the central craters have been formed simultaneously, by the release, with ejective violence, of the matter from its confining crust. The nature of the expansive force being assumed that of solidifying matter, the wide extent of some chasms indicates a deep location of that force. And depth in this matter implies lateness (in the scale of selenological time) of operation, since the central portions of the globe would be the last to cool. Now, we have evidence of comparative lateness afforded by the fact that in many cases the cracks have passed through craters and other asperities which thus obviously existed before the cracking commenced; and thus, so far, the hypothesis of the expansion-cracking is supported by absolute fact.

It may be objected that such an upheaving force as we are invoking, being transitory, would allow the distended surface to collapse again when it ceased to operate, and so close the cracks or chasms it produced. But we consider it not improbable that in some cases, as a consequence of the expansion of subsurface matter, an upflow thereof may have partially filled the crack, and by solidifying have held it open; and it is rational to suppose that there have been various degrees of filling and even of overflow—that in some cases the rising matter has not nearly reached the edge of the crack, as in Fig. 44, while in others it has risen almost to the surface, and in some instances has actually overrun it and produced some sort of elevation along the line of the crack, like that represented sectionally in Fig. 45. It is probable that some of the slightly tumescent lines on the moon's surface have been thus produced.

We have suggested shrinkage as a possible explanation of some cracks. It could hardly have been the direct cause of those compound ones which are distinguished by focal craters, though it may have been a co-operative cause, since the contracting tendency of any area of the crust, by so to

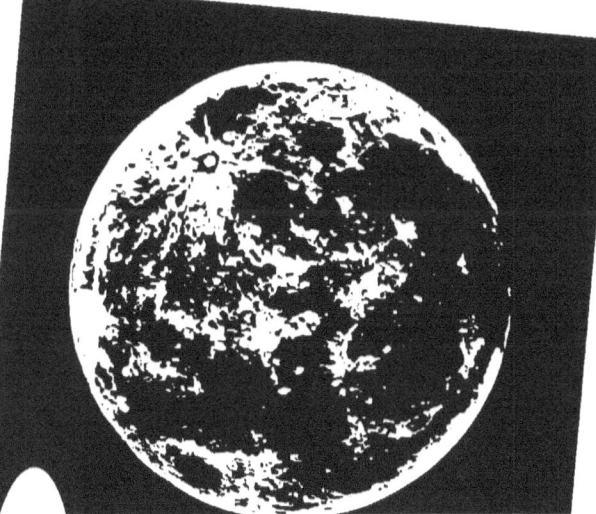

speak weakening it, may have virtually increased the strength of an upheaving force and thus have aided and localized its action. We see, however, no reason why the inevitable ultimate contraction which must

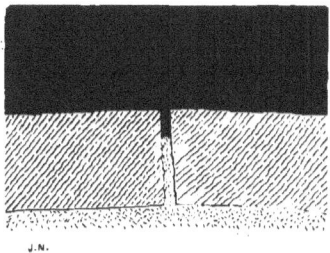

Fig. 44.

have attended the cooling of the moon's crust, even when all internal reactions upon it had ceased, should not have created a class of cracks

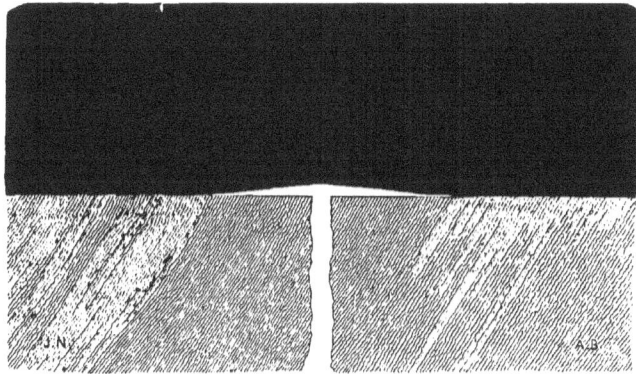

Fig. 45.

without accompanying craters, while it would doubtless have a tendency to increase the length and width of those already existing from any other cause. Some of the more minute clefts, which presumably exist in greater numbers than we yet know of, may doubtless be ascribed to this effect of cooling contraction. In this view we should have to regard such cracks as the latest of all lunar features. Whether the agency that produced them is still at work—whether the cracks are on the increase— is a question impossible of solution : for reasons to be presently adduced, we incline to believe that all cosmical heat passed from the moon, and therefore that it arrived at its present, and apparently final, condition ages upon ages ago.

Besides the ridges spoken of on p. 140, and regarded as cracks up through which matter has been extruded, there are numerous ridges of greater or less extent, which we conceive are of the nature of wrinkles, and have been produced by tangential compression due to the collapse of the moon's crust upon the shrunken interior, as explained and illustrated in Chap. III. The distinguishing feature of the two classes of phenomena we consider to be the presence of a serrated summit in those of the extruded class, while those produced by "wrinkling" action have their summits comparatively free from serration or marked irregularity.

CHAPTER XII.

COLOUR AND BRIGHTNESS OF LUNAR DETAILS; CHRONOLOGY OF
FORMATIONS, AND FINALITY OF EXISTING FEATURES.

SPEAKING generally, the details of the lunar surface seem to us to be devoid of colour. To the naked eye of ordinary sensitiveness the moon appears to possess a silvery whiteness: more critical judges of colour would describe it as presenting a yellowish tinge. Sir John Herschel, during his sojourn at the Cape of Good Hope, had frequent opportunities of comparing the moon's lustre with that of the weathered sandstone surface of Table Mountain, when the moon was setting behind it, and both were illuminated under the same direction of sunlight; and he remarked that the moon was at such times "scarcely distinguishable from the rock in apparent contact with it." Although his observations had reference chiefly to brightness, it can hardly be doubted that similarity of colour is also implied; for any difference in the tint of the two objects would have precluded the use of the words "scarcely distinguishable;" a difference of colour interfering with a comparison of lustre in such an observation, though it must be remembered that he observed through a dense stratum of atmosphere. Viewed in the telescope, the same general yellowish-white colour prevails over all the moon, with a few exceptions offered by the so-called seas. The *Mare Crisium, Mare Serenetatis,* and *Mare Humorum* have somewhat of a greenish tint; the *Palus Somnii* and the circular area of Lichtenberg incline to ruddiness. These tints are, however, extremely faint, and it has been suggested by Arago that they may be mere effects of contrast rather than actual colouration of the surface material. This, however, can hardly be the case, since all the "seas" are not alike affected;

those that are slightly coloured are, as we have said, some green and some red, and contrast could scarcely produce such variations. The supposition of vegetation covering these great flats and giving them a local colour is in our view still more untenable, in the face of the arguments that we shall presently adduce against the possibility of vegetable life existing upon the moon.

It appears to us more rational to consider the tints due to actual colour of the material (presumably lava or some once fluid mineral substance) that has covered these areas; and it may well be conceived that the variety of tint is due to different characters of material, or even various conditions of the same material coming from different depths below the lunar surface; and we may reasonably suppose that the same variously-coloured substances occur in the rougher regions of the lunar surface, but that they exist there in patches too small to be recognized by us, or are "put out" by the brightness to which polyhedral reflexion gives rise.

Seeing that volcanic action has had so large a share in giving to the moon's surface its structural character, analogy of the most legitimate order justifies us in concluding not only that the materials of that surface are of kindred nature to those of the unquestionably volcanic portions of the earth, but also that the tints and colours that characterize terrestrial volcanic and Plutonian products have their counterparts on the moon. Those who have seen the interior and surroundings of a terrestrial volcano after a recent eruption, and before atmospheric agents have exercised their dimming influences, must have been struck with the colours of the erupted materials themselves and the varied brilliant tints conferred on these materials by the sublimated vapours of metals and mineral substances which have been deposited upon them. If, then, analogy is any guide in enabling us to infer the appearance of the invisible from that which we know to be of kindred nature and which we have seen, we may justly conclude that were the moon brought sufficiently near to us to exhibit the minute characteristics of its surface, we should behold the same bright and varied colours in and around its craters that we behold in and about those of the earth; and in all probability the coloured materials of lunar volcanoes would be more fresh and vivid than those of the earth by reason of the absence of those atmospheric elements which tend so rapidly to impair the brightness of coloured surfaces exposed to their influence.

Situated as we are, however, as regards distance from the moon, we have no chance of perceiving these local colours in their smaller masses; but it is by no means improbable, as we have suggested, that the faint tints exhibited by the great plains are due to broad expanses of coloured volcanic material.

But if we fail to perceive diversity of colour upon the lunar surface, we are in a very different position in regard to diversity of brightness or variable light-reflective power of different districts and details. This will be tolerably obvious to those casual observers who have remarked nothing more of the moon's physiography than the resemblance to a somewhat lugubrious human countenance which the full moon exhibits, and which is due to the accidental disposition of certain large and small areas of surface material which have less of the light-reflecting property than other portions; for since all parts seen by a terrestrial observer may be said to be equally shone upon by the sun, it is clear that apparently bright and shaded parts must be produced by differences in the nature of the surface as regards power of reflecting the light received.

When we turn to the telescope and survey the full disc of the moon with even a very moderate amount of optical aid, the meagre impression as to variety of degree of brightness which the unassisted eye conveys is vastly extended and enhanced, for the surface is seen to be diversified by shades of brilliancy and dulness from almost glittering white to sombre grey: and this variety of shading is rendered much more striking by shielding the eye with a dusky glass from the excessive glare, which drowns the details in a flood of light. Under these circumstances the varieties of light and shade become almost bewildering, and defy the power of brush or pencil to reproduce them.

We may, however, realize an imperfect idea of this characteristic of the lunar surface by reference to the self-drawn portrait of the full moon upon Plate III. This is, in fact, a photograph taken from the full moon itself, and enlarged sufficiently to render conspicuous the spots and

U

large and small regions that are strikingly bright in comparison with what may in this place be described as the "ground" of the disc. As an example of a wide and irregularly extensive district of highly reflective material, the region of which Tycho is the central object, is very remarkable. We may refer also to the bright "splashes" of which Copernicus and Kepler are the centres. So brilliant are these spots that they can easily be detected by the unassisted eye about the time of full moon. Still brighter but less conspicuous by its size is the crater Aristarchus, which shines with specular brightness, and almost induces the belief that its interior is composed of some vitreous-surfaced matter: the highly-reflective nature of this object has often caused it to become conspicuous when in the dark hemisphere of the moon, unilluminated by the sun, and lighted only by the light reflected from the earth. At these times it appears so bright that it has been taken for a volcano in actual eruption, and no small amount of popular misconception at one time arose therefrom concerning the conditions of the moon as respects existing volcanic activity—a misconception that still clings to the minds of many.

The parts of the surface distinguished by deficiency of reflecting power are conspicuous enough. We may cite, however, as an example of a detail portion especially remarkable for its dingy aspect, the interior of the crater Plato, which is one of the darkest spots (the darkest well-defined one) upon the hemisphere of the moon visible to us. For facilitating reference to shades of luminosity, Schroeter and Lohrman assorted the variously reflective parts into 10 grades, commencing with the darkest. Grades 1 to 3 comprised the various deep greys; 4 and 5 the light greys; 6 and 7 white; and 8 to 10 brilliant white. The spots Grimaldi and Riccioli came under class 1 of this notation; Plato between 1 and 2. The "seas" generally ranged from 2 to 3; the brightest mountainous portions mostly between degrees 4 and 6; the crater walls and the bright streaks came between these and the bright peaks, which fell under the 9th grade. The maximum brightness, the 10th grade, is instanced only in the case of Aristarchus and a point in Werner, though Proclus nearly approaches it, as do many bright spots, chiefly the sites of minute craters, which make their appearance at the time of full moon.

In photographic pictures produced by the moon of itself there is always an apparent exaggeration in the relation of light to dark portions of the disc. The dusky parts look, upon the photograph, much darker than to the eye directed to the moon itself, whether assisted or not by optical appliances. It may be that the real cause of this discrepancy is that the eye fails to discover the actual difference upon the moon itself, being insensible to the higher degrees of brightness or not estimating them at their proper brilliance with respect to parts less bright. On the other hand, it is probable that the enhanced contrast in the photograph is due to some peculiar condition of the darker surface matter affecting its power of reflecting the actinic constituent of the rays that fall upon it.

The study of the varying brightness or reflective power of different regions and spots of the lunar disc leads us to the consideration of the relative antiquity of the surface features; for it is hardly possible to regard these variations attentively without being impressed with the conviction that they have relation to some chronological order of formation. We cannot, in the first place, resist the conviction that the brightest features were the latest formed; this strikes us as evident on *primâ facie* grounds; but it becomes more clearly so when we remark that the bright formations, as a rule, overlie the duller features. The elevated parts of the crust are brighter than the "seas" and other areas; and it is pretty clear that the former are newer than the latter, upon which they appear to be super-imposed, or through which they seem to have extruded.* The vast dusky plains are in every instance more or less sprinkled with spots and minute craters, and these last were obviously formed after the area that contains them. One is almost disposed to place the order of formations in the order of relative brightness, and so consider the dingiest parts the oldest and the brightest spots and craters the newest features, though, in the absence of an atmosphere competent to impair the reflective power of the surface materials, we are unable to justify this classification by suggesting a cause for such a deterioration by time as the hypothesis pre-supposes.

As we have entered upon the question of relative age of the lunar

---

* We meet a difficulty in reconciling this idea with the partial craters of which we have a conspicuous example in Fracastorius, No. 78, of our Map, which seem to be partially sunk below the contiguous surface. This looks as though the crater-rim belonged to an older epoch than the plain from which it rises.

features, we may remark that there are evidences of various epochs of formation of particular classes of details, irrespective of their condition in respect of brightness, or, as we may say, freshness of material. As a rule, the large craters are older than the small ones. This is proved by the fact that a large object of this class is never seen to interfere with or overlap a small one. Those of nearly equal size are, however, seen to overlap one another as though several eruptions of equal intensity had occurred from the same source at different points. This is strikingly instanced in the group of craters situated in the position 35—141 on our map, the order of formation of each of which is clearly apparent. The region about Tycho offers an inexhaustible field for study of these phenomena of overlapping or interpolating craters, and it will be found, with very few exceptions, that the smaller crater is the impinging or parasitical one, and must therefore have been formed after the larger, upon which it intrudes or impinges. There are frequent cases in which a large crater has had its rampart interrupted by a lesser one, and this again has been broken into by one still smaller; and instances may be found where a fourth crater smaller than all has intruded itself upon the previous intruder. The general tendency of these examples is to show that the craters diminished in size as the moon's volcanic energy subsided: that the largest were produced in the throes of its early violence, and that the smallest are the results of expiring efforts possibly impeded through the deep-seatedness of the ejective source.

Another general fact of this chronological order is that the mountain chains are never seen to intrude upon formations of the crater order. We do not anywhere find that a mountain chain runs absolutely into or through a crater; but, on the other hand, we do find that craters have formed on mountain chains. This leads unmistakably to the inference that the craters were not formed *before* their allied mountain chains; and we might assume therefore that the mountains generally are the older formations, but that there is nothing to prove that the two classes of features, where they intermingle, as in the Apennines and Caucasus, were not erupted cotemporaneously.

Upon the assumption that the latest ejected or extruded matter is that which is brightest, we should place the bright streaks among the more

OVERLAPPING CRATERS

recent features. Be this as it may, it is tolerably certain that the cracks, whose apparently close relation to the radiating streaks we have endeavoured to point out, are relatively of a very late formative period. We are indeed disposed to consider them as the most recent features of all : the evidence in support of this consideration being the fact that they are sometimes found intersecting small craters that, from the way in which they are cut through by the cracks, must have been *in situ* before the cracking agency came into operation. It is in accordance with our hypothesis of the moon's transition from a fluid to a solid body to consider that a cracking of the surface would be the latest of all the phenomena produced by contraction in final cooling.

The foregoing remarks naturally lead us to the question whether changes are still going on upon the surface of our satellite : whether there is still left in it a spark of its volcanic activity, or whether that activity has become totally extinct. We shall consider this question from the observational and theoretical point of view. First as regards observations. This much may be affirmed indisputably—that no object or detail visible to the earliest selenographers (whose period may be dated 200 years back) has altered from the date of their maps to the present. When we pass from the bolder features to the more minute details we find ourselves at a loss for materials for forming an inference ; the only map pretending to accuracy even of the larger among small objects being that of Beer and Maedler, which, truly admirable as it is, is not very safely to be relied upon for settling any question of alleged change, on account of the conventional system adopted for exhibiting the forms of objects, every object being mapped rather than drawn, and shown as it never is or can be presented to view on the moon itself. This difficulty would present itself if a question of change were ever raised upon the evidence of Beer and Maedler's map : it may indeed have prevented such a question being raised, for certainly no one has hitherto been bold enough to assert that any portion or detail of the map fails to represent the actual state of the moon at the present time.

In default of published maps, we are thrown for evidence on this question upon observations and recollections of individual observers whose familiarity with the lunar details extends over lengthy periods. Speaking for ourselves, and upon the strength of close scrutinies continued with assiduity through the past thirty years, we may say that we have never had the suspicion suggested to our eye of any actual change whatever having taken place in any feature or minute detail of the lunar surface; and our scrutinies have throughout been made with ample optical means, mostly with a 20-inch reflector. This experience has made us not unnaturally in some slight degree sceptical concerning the changes alleged to have been detected by others. Those asserted by Schroeter and Gruithuisen were long ago rejected by Beer and Maedler, who explained them, where the accuracy of the observer was not questioned, by variations of illumination, a cause of illusory change which is not always sufficiently taken into account. A notable instance of this deception occurred a few years ago in the case of the minute bright crater *Linné*, which was for a considerable period declared, upon the strength of observations of very promiscuous character, to be varying in form and dimensions almost daily, but the alleged constant changes of which have since been tacitly regarded as due to varying circumstances of illumination induced by combinations of libratory effects with the ordinary changes depending upon the direction of the sun's rays as due to the age of the moon. This explanation does not, however, dispose of the question whether the crater under notice suffered any actual change before the hue and cry was raised concerning it. Attention was first directed to it by Schmidt, of Athens, whose powers of observation are known to be remarkable, and whose labours upon the moon are of such extent and minuteness as to claim for his assertions the most respectful consideration.* He affirmed in 1866 that the crater at that date presented an appearance decidedly different from that which it had had since 1841: that whereas it had been from the earlier epoch always easily seen as a very deep crater, in October 1866 and thenceforward it presented only a white spot, with at most but a very shallow aperture, very difficult to be detected. Schmidt is one of the very few observers whose

---

* We are informed by a friend, who has lately visited Athens, that Schmidt's detail drawings of the Moon, comprising the work of forty years, form a small library in themselves. The map embodying them is so large (6 ft. 6 in. in diameter) and so full of detail that there is small hope of its complete publication, unless there should be such a wide extension of interest in the minute study of our satellite as to justify the cost of reproducing it.

long familiarity with the moon entitles him to speak with confidence upon such a question as that before us upon the sole strength of his own experience; and this case is but an isolated one, at least it is the only one he has brought forward. He is, however, still firmly convinced that it is an instance of actual change, and not an illusion resulting from some peculiar condition of illumination of the object. It should be added also on this side of the discussion that an English observer, the Rev. T. W. Webb, while apparently indisposed to concede the supposition of any notable changes in the lunar features, has yet found from his own observations that, after all due allowance for differences of light and shade upon objects at different times, there is still a "residuum of minute variations not thus disposed of" which seem to indicate that eruptive action in the moon has not yet entirely died out, though its manifestation at present is very limited in extent. It appears to us that, if evidence of continuing volcanic action is to be sought on the moon, the place to look for it is around the circumference of the disc, where eruption from any marginal orifice would manifest itself in the form of a protruding haziness, somewhat as illustrated to an exaggerated extent in the annexed cut.

FIG. 46.

The theoretical view of the question, which we have now to consider, has led us, however, to the strong belief that no vestige of its former volcanic activity lingers in the moon—that it assumed its final condition an inconceivable number of ages ago, and that the high interest which would attach to the close scrutiny of our satellite if it *were* still the theatre of volcanic reaction cannot be hoped for. If it be just and allowable to assume that the earth and the moon were condensed into planetary form at nearly the same epoch (and the only rational scheme of cosmogony justifies the assumption) then we may institute a comparison between the condition of the two bodies as respects their volcanic age, using the one as a basis for inference concerning the state of the other. We have reason to believe that the earth's crust has nearly assumed its final state so far as volcanic reactions of its interior upon its exterior are concerned: we may affirm that within the historical period no igneous convulsions of any considerable magnitude have occurred; and we may consider that the volcanoes now active over the surface of the globe represent the last expiring efforts of its eruptive force. Now in the earth we perceive several conditions wherefrom we may infer that it parted with its cosmical heat (and therefore with its prime source of volcanic agency) at a rate which will appear relatively very slow when we come to compare the like conditions in the moon. We may, we think, take for granted that the surface of a planetary body generally determines its *heat dispersing* power, while its volume determines its *heat retaining* power. Given two spherical bodies of similar material but of unequal magnitude and originally possessing the same degree of heat, the smaller body will cool more rapidly than the larger, by reason of the greater proportion which the surface of the smaller sphere bears to its volume than that of the larger sphere to its volume—this proportion depending upon the geometrical ratio which the surfaces of spheres bear to their volumes, the contents of spheres being as the *cubes* and the surfaces as the *squares* of their diameters. The volume of the earth is 49 times as great as that of the moon, but its surface is only 13 times as great; there is consequently in the earth a power of retaining its cosmical heat nearly four times as great as in the case of the moon; in other words, the moon and earth being supposed at one time to have had an equally high temperature, the moon would cool down to a given low temperature in about one fourth the time that the earth would require to cool to the same temperature.

But the earth's cosmical heat has without doubt been considerably conserved by its vaporous atmosphere, and still more by the ocean in its antecedent vaporous form. Yet notwithstanding all this, the earth's surface has nearly assumed its final condition so far as volcanic agencies are concerned : it has so far cooled as to be subject to no considerable distortions or disruptions of its surface. What then must be the state of the moon, which, from its small volume and large proportionate area, parted with its heat at the above comparatively rapid rate? The matter of the moon is, too, less dense than the earth, and hence doubtless from this cause disposed to more rapid cooling ; and it has no atmosphere or vaporous envelope to retard its radiating heat. We are driven thus to the conclusion that the moon's loss of cosmical heat must have been so rapid as to have allowed its surface to assume its final conformation ages on ages ago, and hence that it is unreasonable and hopeless to look for evidence of change of any volcanic character still going on.

We conceive it possible, however, that minute changes of a non-volcanic character may be proceeding in the moon, arising from the violent alternations of temperature to which the surface is exposed during a lunar day and night. The sun, as we know, pours down its heat unintermittingly for a period of fully 300 hours upon the lunar surface, and the experimental investigations of Lord Rosse, essentially confirmed by those of the French observer, Marie Davy, show that under this powerful insolation the surface becomes heated to a degree which is estimated at about 500° of Fahrenheit's scale, the fusing point of tin or bismuth. This heat, however, is entirely radiated away during the equally long lunar night, and, as Sir John Herschel surmised, the surface probably cools down again to a temperature as low as that of interstellar space : this has been assumed as representing the absolute zero of temperature which has been calculated from experiments to be 250° below the zero of Fahrenheit's scale. Now such a severe range of heat and cold can hardly be without effect upon some of the component materials of the lunar surface.* If there be any such materials as the vitreous lavas that are

---

* It is conceivable that the alleged changes in the crater Linné may have been caused by a filling of the crater by some such crumbling action as we are here contemplating.

found about our volcanoes, such as obsidian for instance, they are doubtless cracked and shivered by these extreme transitions of temperature; and this comparatively rapid succession of changes continued through long ages would, we may suppose, result in a disintegration of some parts of the surface and at length somewhat modify the selenographic contour. It is, however, possible that the surface matter is mainly composed of more crystalline and porous lavas, and these might withstand the fierce extremes like the "fire-brick" of mundane manufacture, to which in molecular structure they may be considered comparable. Lavas as a rule are (upon the earth) of this unvitreous nature, and if they are of like constitution on the moon, there will be little reason to suspect changes from the cause we are considering. Where, however, the material, whatever its nature, is piled in more or less detached masses, there will doubtless be a grating and fracturing at the points of contact of one mass with another, produced by alternate expansions and contractions of the entire masses, which in the long run of ages must bring about dislocations or dislodgments of matter that might considerably affect the surface features from a close point of view, but which can hardly be of sufficient magnitude to be detected by a terrestrial observer whose best aids to vision give him no perception of minute configurations. And it must always be borne in mind that changes can only be *proved* by reference to previous observations and delineations of unquestionable accuracy.

Speaking by our own lights, from our own experience and reasoning, we are disposed to conclude that in all visible aspects the lunar surface is unchangeable, that in fact it arrived at its terminal condition *eons* of ages ago, and that in the survey of its wonderful features, even in the smallest details, we are presented with the sight of objects of such transcendent antiquity as to render the oldest geological features of the earth modern by comparison.

CHAPTER XIII.

THE MOON AS A WORLD: DAY AND NIGHT UPON ITS SURFACE.

A WIDE interest, if not a deep one, attaches to the general question as to the existence of living beings, or at least the possibility of organic existence, on planetary bodies other than our own. The question has been examined in all ages, by the lights of the science peculiar to each. With every important accession to our astronomical knowledge it has been re-raised: every considerable discovery has given rise to some new step or phase in the discussion, and in this way there has grown up a somewhat extensive literature exclusively relating to mundane plurality. It will readily be understood that the moon, from its proximity to the earth, has from the first received a large, perhaps the largest, share of attention from wanderers in this field of speculation: and we might add greatly to the bulk of this volume by merely reviewing some of the more curious and, in their way, instructive conjectures specially relating to the moon as a world—to imaginary journeys towards her, and to the beings conjectured to dwell upon and within her. This, however, we feel there is no occasion to do, for it is our purpose merely to point out the two or three almost conclusive arguments against the possibility of any life, animal or vegetable, having existence on our satellite.

We well know what are the requisite conditions of life on the earth; and we can go no further for grounds of inference; for if we were to start by assuming forms of life capable of existence under conditions widely and essentially different from those pertaining to our planet, there would be no need for discussing our subject further: we could revel in conjectures, without a thought as to their extravagance. The only legitimate phase of the question we can entertain is this:—can there be on the moon any kind of living things analogous to any kind of living things upon the earth? And this question, we think, admits only of a negative answer. The lowest forms of vitality cannot exist without air, moisture, and a moderate range of temperature. It may be true, as recent experiments seem to show, that organic germs will retain their vitality without either of the first, and with exposure to intense cold and to a considerable degree of heat; and it is conceivable that the mere germs of life may be present on the moon.* But this is not the case with living organisms themselves. We have, in Chapter V., specially devoted to the subject, cited the evidence from which we know that there can be at the most, no more air on the moon than is left in the receiver of an air-pump after the ordinary process of exhaustion. And with regard to moisture, it could not exist in any but the vaporous state, and we know that no appreciable amount of vapour can be discovered by any observation (and some of them are crucial enough) that we are capable of making. We may suppose it just within the verge of possibility that some low forms of vegetation might exist upon the moon with a paucity of air and moisture such as would be beyond even our most severe powers of detection: but granting even this, we are met by the temperature difficulty; for it is inconceivable that any plant-life could survive exposure first to a degree of cold vastly surpassing that of our arctic regions, and then in a short time (14 days) to a degree of heat capable of melting the more fusible metals—the total range being equal, as we have elsewhere shown, to perhaps 600 or 700 degrees of our thermometric scale.

The higher forms of vegetation could not reasonably be expected to exist under conditions which the lower forms could not survive. And as regards the possibility of the existence of animal life in any form or condition on the lunar surface, the reasons we have adduced in reference

* Is it not conceivable that the protogerms of life pervade the whole universe, and have been located upon every planetary body therein? Sir William Thomson's suggestion that life came to the earth upon a seed-bearing meteor was weak, in so far that it shifted the locus of life-generation from one planetary body to another. Is it not more philosophical, more consistent with our conception of Creative omnipotence and impartiality, to suppose that the protogerms of life have been sown broadcast over all space, and that they have fallen here upon a planet under conditions favourable to their development, and have sprung into vitality when the fit circumstances have arrived, and there upon a planet that is, and that may be for ever, unfitted for their vivification.

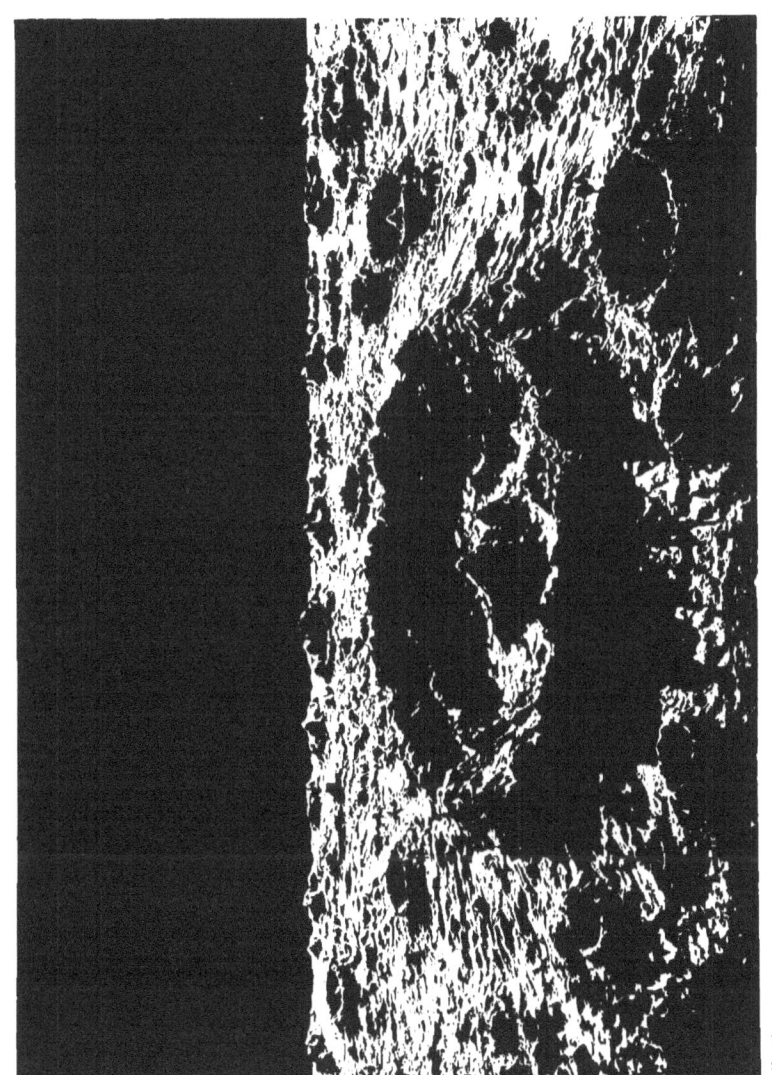

to the non-existence of vegetable life bear still more strongly against the possibility of the existence of the former. We know of no animal that could live in what may be considered a vacuum and under such thermal conditions as we have indicated.

As to man, aëronautic experience teaches us that human life is endangered when the atmosphere is still sufficiently dense to support 12 inches of mercury in the barometer tube; what then would be his condition in a medium only sufficiently dense to sustain one-tenth of an inch of the barometric column? We have evidence from the most delicate tests that no atmosphere or vapour approaching even this degree of attenuation exists around the moon's surface.

Taking all these adverse conditions into consideration, we are in every respect justified in concluding that there is no possibility of animal or vegetable life existing on the moon, and that our satellite must therefore be regarded as a barren world.

* * * * * *

After this disquisition upon lunar uninhabitability it may appear somewhat inconsistent for us to attempt a description of the scenery of the moon and some other effects that would be visible to a spectator, and of which he would be otherwise sensible, during a day and a night upon her surface. But we can offer the sufficient apology that an imaginary sojourn of one complete lunar day and night upon the moon affords an opportunity of marshalling before our readers some phenomena that are proper to be noticed in a work of this character, and that have necessarily been passed over in the series of chapters on consecutive and special points that have gone before. It may be urged that, in depicting the moon from such a standpoint as that now to be taken, we are describing scenes that never have been such in the literal sense of the word, since no eye has ever beheld them. Still we have this justification—that we are invoking the conception of things that actually exist; and that we are not, like some imaginary voyagers to the moon, indulging in mere flights of fancy. Although it is impossible for a habitant of this earth fully to realize existence upon the moon, it is yet possible, indeed almost inevitable, for a thoughtful telescopist—watching the moon night after night, observing the sun rise upon a lunar scene, and noting the course of effects that follow till it sets—it is almost inevitable, we say, for such an observer to identify himself so far with the object of his scrutiny, as sometimes to become in thought a lunar being. Seated in silence and in solitude at a powerful telescope, abstracted from terrestrial influences, and gazing upon the revealed details of some strikingly characteristic region of the moon, it requires but a small effort of the imagination to suppose one's self actually upon the lunar globe, viewing some distant landscape thereupon ; and under these circumstances there is an irresistible tendency in the mind to pass beyond the actually *visible*, and to fill in with what it knows must exist those accessory features and phenomena that are only hidden from us by distance and by our peculiar point of view. Where the material eye is baffled, the clairvoyance of reason and analogy comes to its aid.

Let us then endeavour to realize the strange consequences which the position and conditions of the moon produce upon the aspect of a lunar landscape in the course of a lunar day and night.

The moon's day is a long one. From the time that the sun rises upon a scene* till it sets, a period of 304 hours elapses, and of course double this interval passes between one sunrise and the next. The consequences of this slow march of the sun begin to show themselves from the instant that he rises above the lunar horizon. Dawn, as we have it on earth, can have no counterpart upon the moon. No atmosphere is there to reflect the solar beams while the luminary is yet out of actual sight, and only the glimmer of the zodiacal light heralds the approach of day. From the black horizon the sun suddenly darts his bright untempered beams upon the mountain tops, crowning them with dazzling brilliance while their flanks and valleys are yet in utter darkness. There is no blending of the night into day. And yet there is a growth of illumination that in its early stages may be called a twilight, and which is caused by the slow rise of the sun. Upon the earth, in central latitudes, the average time occupied by the sun in rising, from the first glint of his upper edge till the whole disc is in sight, is but two minutes and a quarter. Upon the moon,

* Our remarks have general reference to a region of the moon near her equator; near the poles some of the conditions we shall describe would be somewhat modified.

however, this time is extended to a few minutes short of an hour, and therefore, during the first few minutes a dim light will be shed by the small visible chord of the solar disc, and this will give a proportionately modified degree of illumination upon the prominent portion of the landscape, and impart to it something of the weird aspect which so strikes an observer of a total solar eclipse on earth when the scene is lit by the thin crescent of the re-appearing sun. This impaired illumination constitutes the only dawn that a lunar spectator could behold. And it must be of short duration; for when, in the course of half an hour, the solar disc has risen half into view the lighting would no doubt appear nearly as bright to the eye as when the entire disc of the sun is above the horizon. In this lunar sunrise, however, there is none of that gilding and glowing which makes the phenomenon on earth so gorgeous. Those crimson sky-tints with which we are familiar are due to the absorption of certain of the polychromous rays of light by our atmosphere. The blue and violet components of the solar beams are intercepted by our envelope of vapour, and only the red portions are free to pass; while on the moon, as there is no atmosphere, this selective absorption does not occur. If it did, an observer gazing from the earth upon the regions of the moon upon which the sun is just rising would see the surface tinted with rosy light. This, however, is not the case: the faintest lunar features just catching the sun are seen simply under white light diluted to a low degree of brightness. Only upon rare occasions is the lunar scenery suffused with coloured illumination, and these are when, as we shall presently have to describe, the solar rays reach the moon after traversing the earth's atmosphere during an eclipse of the sun.

This atmosphere of ours is the most influential element in beautifying our terrestrial scenery, and the absence of such an appendage from the moon is the great modifying cause that affects lunar scenery as compared with that of the earth. We are accustomed to the sun with its dazzling brightness—overpowering though it be—subdued and softened by our vaporous screen. Upon the moon there is no such modification. The sun's intrinsic brilliancy is undiminished, its apparent distance is shortened, and it gleams out in fierce splendour only to be realized, and then imperfectly, by the conception of a gigantic electric light a few feet from the eye. And the brightness is rendered the more striking by the blackness of the surrounding sky. Since there is no atmosphere there can be no sky-light, for there is nothing above the lunar world to diffuse the solar beams; not a trace of that moisture which even in our tropical skies scatters some of the sun's light and gives a certain degree of opacity or blueness, deep though it be, to the heavens by day. Upon the moon, with no light-diffusing vapour, the sky must be as dark or even darker than that with which we are familiar upon the finest of moonless nights. And this blackness prevails in the full blaze of the lunar noon-day sun. If the eye (upon the moon) could bear to gaze upon the solar orb (which would be less possible than upon earth) or could it be screened from the direct beams, as doubtless it could by intervening objects, it would perceive the nebulous and other appendages which we know as the corona, the zodiacal light, and the red solar protuberances: or if these appendages could not be viewed with the sun above the horizon they would certainly be seen in glorious perfection when the luminary was about to rise or immediately after it had set.

And, notwithstanding the sun's presence, the planets and stars would be seen to shine more brilliantly than we see them on the clearest of nights; the constellations would have the same configurations, though they would be differently situated with respect to the celestial pole about which they would appear to turn, for the axis of rotation of the moon is directed towards a point in the constellation Draco. The stars would never twinkle or change colour as they appear to us to do, for scintillation or twinkling is a phenomenon of atmospheric origin, and they would retain their full brightness, down even to the horizon, since there would be no haze to diminish their light. The planets, and the brighter stars at least, would be seen even when they were situated very near to the sun. The planet Mercury, so seldom detected by terrestrial gazers, would be almost constantly in view during the lunar day, manifesting his close attendance on the central luminary by making only short excursions of about two (lunar) days' length, first on one side and then on the other. Venus would be nearly as continuously visible, though her wanderings would be more extensive on either side. The zodiacal light also, which in our English latitude and climate is but rarely seen and in more favourable climes appears only when the sun itself is hidden beneath the horizon, would upon the moon be seen as a constant accompaniment to the luminary throughout his daily course across the lunar sky. The other planets would appear generally as they do to us on earth, but, never being lost in daylight, their courses among the stars could be traced with scarcely any interruption.

One planet, however, that adorns the sky of the lunar hemisphere which is turned towards us deserves special mention from the conspicuous and highly interesting appearance it must present. We allude to the earth. To nearly one-half of the moon (that which we never see) this imposing object can never be visible; but to the half that faces us the terrestrial planet must appear almost fixed in the sky. A lunar spectator in (what is to us) the centre of the disc, or about the region north of the lunar mountains Ptolemy and Hipparchus, would have the earth in his zenith. From regions upon the moon a little out of what is to us the centre, a spectator would see the earth a little declining from the zenith, and this declination would increase as the regions corresponding to the (to us) apparent edge of the moon were approached, till at the actual edge it would be seen only upon the horizon. From the phenomena of libration (explained in Chap. VI.) the earth would appear from nearly all parts of the lunar hemisphere to which it is visible at all to describe a small circle in the sky. To an observer, however, upon the (to us) marginal regions of the lunar globe, it would appear only during a portion of the lunar day—being visible in fact only in that part of its small circular path which happened to lie above the observer's horizon: in some regions only a portion of the terrestrial disc would make its brief appearance. From the lunar hemisphere beyond this marginal line the earth can never be seen at all.

The lunar spectator whose situation enabled him to view the earth would see it as a moon; and a glorious moon indeed it must be. Its diameter would be four times as great as that of the moon itself as seen by us, and the area of its full disc 13 times as great. It would be seen to pass through its phases, just as does our satellite, once in a lunar day or a terrestrial month, and during that cycle of phases, since 29 of our days would be occupied by it, the axial rotation would bring all the features of its surface configuration into view so many times in succession. But the greatest beauty of this noble moon would be seen during the lunar night, in considering which we shall again allude to it; for when it is full-moon to the earth it is new-earth to the moon. At lunar midnight this globe of ours is fully illuminated; as morning nears, the earth-moon wanes, its disc slowly passing through the gibbous phases until at sunrise it would be just half-illuminated. During the long forenoon it assumes a crescent which narrows and narrows till at midday the sun is in line with the earth and the latter is invisible, save perhaps by a thin line of light marking its upper or lower edge, accordingly as the sun is apparently above or below it. In the lunar afternoon an illuminated crescent appears upon the opposite side of the terrestrial globe, and this widens and widens till it becomes a half disc by lunar sunset and a full disc by lunar midnight.

The sun in his daily course passes at various distances, sometimes above and sometimes below, the nearly stationary earth. Obviously it will at times pass actually behind it, and then the lunar spectator would behold the sublime spectacle of a total solar eclipse, and that under circumstances which render the phenomenon far more imposing than its counterpart can appear from the earth; for whereas, when we see the moon eclipse the sun, the nearly similar (apparent) diameters of the two bodies renders the duration of totality extremely short—at most 7 minutes —a lunar spectator, the earth appearing to him four times the diameter of the sun, and he and the earth being relatively stationary, would enjoy a view of the totality extending over several hours. During the passage of the solar disc behind that of the earth, a beautiful succession of luminous phenomena would be observed to follow from the refractions and dispersions which the sunbeams would suffer in passing tangentially through those parts of our atmospheric envelope which lie in their course; those, for instance, on the margin of the earth, as seen from the moon. As the sun passed behind the earth, the latter would be encircled upon the in-going side with a beautiful line of golden light, deepening in places to glowing crimson, due to the absorption, already spoken of, of all but the red and orange rays of the sun's light by the vapours of our atmosphere. As the eclipse proceeded and totality came on, this ruddy glow would extend itself nearly, if not all, around the black earth, and so bright would it be, that the whole lunar landscape covered by the earth's shadow would be illuminated with faint crimson light,* save, perhaps, in some parts of the far distance, upon which the earth had not yet cast its shadow, or off which the shadow had passed. Although the crimson light would preponderate, it would not appear bright and red alike all around the earth's periphery. The circle of light would be, in fact, *the ring of twilight* round our globe, and it would only appear red in those places where the atmosphere chanced to be in that condition favourable for producing what on earth we know as red sunset and sunrise. We know that the sun, even in clear sky, does not always set and rise with the beautiful red glow, which may be determined by merely local causes, and will therefore vary in different parts of the earth. Now a lunar spectator watching the sun eclipsed by the earth, would see, during totality and at a *coup d'œil*, every point around our world upon which the sun is setting on one side and rising upon the other. To every part of the earth around what is then the margin, as seen from the moon, the sun is upon the horizon, shining through a great thickness of atmosphere, reddening it, and being reddened by it wherever the vaporous conditions conduce to that colouration. And at all parts where these conditions obtain, the lunar eclipse-observer would see the ring of light around the black earth-globe brilliantly crimsoned; at other parts it would have other shades of red and yellow, and the whole effect would be to make the grand earth-ball, hanging in the lunar sky, like a dark sphere in a circle of glittering gold and rubies.

During the early stages of the eclipse, this chaplet of brilliant-coloured lights would be brightest upon the side of the *disappearing* sun; at the time of central eclipse the radiance (supposing the sun to pass centrally behind the earth) would be equally distributed, and during the later stages it would preponderate upon the side of the *reappearing* sun. We

---

* We see this reddening during an eclipse of the moon (when the event we are describing—an eclipse of the sun visible from the moon—really takes place). The blood-red colour has often struck observers very forcibly, and it has indeed been suggested that the appearance may be the innocent and oft-repeated fulfilment of the prophetic allusion to the moon being "turned into blood."

have endeavoured to give a pictorial realization of this phenomenon and of the effect of the eclipse upon the lunar landscape, but such a picture cannot but fall very, very far short of the reality. (See Plate XXII.)

And now for a time let us turn attention from the lunar sky to the scenery of the lunar landscape. Let us, in imagination, take our stand high upon the eastern side of the rampart of one of the great craters. Height, it must be remarked, is more essential on the moon to command extent of view than upon the earth, for on account of the comparative smallness of the lunar sphere the dip of the horizon is very rapid. Such height, however, would be attained without great exercise of muscular power, since equal amounts of climbing energy would, from the smallness of lunar gravity, take a man six times as high on the moon as on the earth. Let us choose, for instance, the hill-side of Copernicus. The day begins by a sudden transition. The faint looming of objects under the united illumination of the half-full earth, and the zodiacal light is the lunar precursor of daybreak. Suddenly the highest mountain peaks receive the direct rays of a portion of the sun's disc as it emerges from below the horizon. The brilliant lighting of these summits serves but to increase, by contrast, the prevailing darkness, for they seem to float like islands of light in a sea of gloom. At a rate of motion twenty-eight times slower than we are accustomed to, the light tardily creeps down the mountain-sides, and in the course of about twelve hours the whole of the circular rampart of the great crater below us, and towards the east, shines out in brilliant light, unsoftened by a trace of mountain-mist. But on the opposite side, looking into the crater, nothing but blackness is to be seen. As hour succeeds hour, the sunbeams reach peak after peak of the circular rampart in slow succession, till at length the circle is complete and the vast crater-rim, 50 miles in diameter, glistens like a silver-margined abyss of darkness. By-and-by, in the centre, appears a group of bright peaks or bosses. These are the now illuminated summits of the central cones, and the development of the great mountain cluster they form henceforth becomes an imposing feature of the scene. From our high standpoint, and looking backwards to the sunny side of our cosmorama, we glance over a vast region of the wildest volcanic desolation. Craters from five

miles diameter downwards crowd together in countless numbers, so that the surface, as far as the eye can reach, looks veritably frothed over with them. Nearer the base of the rampart on which we stand, extensive mountain chains run to north and to south, casting long shadows towards us; and away to southward run several great chasms a mile wide and of appalling blackness and depth. Nearer still, almost beneath us, crag rises on crag and precipice upon precipice, mingled with craters and yawning pits, towering pinnacles of rock and piles of scoria and volcanic *débris*. But we behold no sign of existing or vestige of past organic life. No heaths or mosses soften the sharp edges and hard surfaces: no tints of cryptogamous or lichenous vegetation give a complexion of life to the hard fire-worn countenance of the scene. The whole landscape, as far as the eye can reach, is a realization of a fearful dream of desolation and lifelessness—not a dream of death, for that implies evidence of pre-existing life, but a vision of a world upon which the light of life has never dawned.

Looking again, after some hours' interval, into the great crateral amphitheatre, we see that the rays of the morning sun have crept down the distant side of the rampart, opposite to that on which we stand, and lighted up its vast landslipped terraces into a series of seeming hill-circles with all the rude and rugged features of a terrestrial mountain view, and none of the beauties save those of desolate grandeur. The plateau of the crater is half in shadow 10,000 feet below, with its grand group of cones, now fully in sight, rising from its centre. Although these last are twenty miles away and the base of the opposite rampart fully double that distance, we have no means of judging their remoteness, for in the absence of an atmosphere there can be no aërial perspective, and distant objects appear as brilliant and distinct as those which are close to the observer. Not the brightness only, but the various colours also of the distant objects are preserved in their full intensity; for colour we may fairly assume there must be. Mineral chlorates and sublimates will give vivid tints to certain parts of the landscape surface, and there must be all the more sombre colours which are common to mineral matters that have been subjected to fiery influence. All these tints will shine and glow with their greater or less intrinsic lustres, since they have not been deteriorated by atmospheric agencies, and far and near they will appear clear alike, since there is no aërial medium to veil them or tarnish their pristine brightness.

In the lunar landscape, in the line of sight, there are no means of estimating distances; only from an eminence, where the intervening ground can be seen, is it possible to realize *magnitude* in a lunar cosmorama and comprehend the dimensions of the objects it includes.

And with no air there can be no diffusion of light. As a consequence, no illumination reaches those parts of the scene which do not receive the direct solar rays, save the feeble amount reflected from contiguous illuminated objects, and a small quantity shed by the crescent earth. The shadows have an awful blackness. As we stand upon our chosen point of observation, we see on the lighted side of the rampart almost dazzling brightness, while beneath us, on the side away from the sun, there is a region many miles in area impenetrable to the sight, for there is no object within it receiving sufficient light to render it discernible; and all around us, far and near, there is the violent contrast between intense brightness of insulated parts and deep gloom of those in equally intense shadow. The black though starlit sky helps the violence of this contrast, for the bright mountains in the distance around us stand forth upon a background formed by the darkness of interplanetary space. The visible effects of these conditions must be in every sense unearthly and truly terrible. The hard, harsh glowing light and pitchy shadows; the absence of all the conditions that give tenderness to an earthly landscape; the black noonday sky, with the glaring sun ghastly in its brightness; the entire absence of vestiges of any life save that of the long since expired volcanoes—all these conspire to make up a scene of dreary, desolate grandeur that is scarcely conceivable by an earthly habitant, and that the description we have attempted but insufficiently pourtrays.

A legitimate extension of the imagination leads us to impressions of lunar conditions upon other senses than that of sight, to which we have hitherto confined our fancy. We are met at the outset with a difficulty in this extension; for it is impossible to conceive the sensations which the absence of an atmosphere would produce upon the most important of our bodily functions. If we would attempt the task we must conjure up feelings of suffocation, of which the thoughts are, however, too horrible to be dwelt upon; we must therefore maintain the delusion that we can exist without air, and attempt to realize some of the less discomforting effects of the absence of this medium. Most notable among these are the untempered heat of the direct solar rays, and the influence thereof upon the surface material upon which we suppose ourselves to stand. During a a period of over three hundred hours the sun pours down his beams with unmitigated ferocity upon a soil never sheltered by a cloud or cooled by a shower, till that soil is heated, as we have shown, to a temperature equal nearly to that of melting lead; and this scorching influence is felt by everything upon which the sun shines on the lunar globe. But while regions directly isolated are thus heated, those parts turned from the sun would remain intensely cold, and that scorching in sunshine and freezing in shade with which mountaineers on the earth are familiar would be experienced in a terribly exaggerated degree. Among the consequences, already alluded to, of the alternations of temperature to which the moon's crust is thus exposed, are doubtless more or less considerable expansions and contractions of the surface material, and we may conceive that a cracking and crumbling of the more brittle constituents would ensue, together with a grating of contiguous but disconnected masses, and an occasional dislocation of them. We refer again to these phenomena to remark that if an atmospheric medium existed they would be attended with noisy manifestations. There are abundant causes for grating and crackling sounds, and such are the only sources of noise upon the moon, where there is no life to raise a hum, no wind to murmur, no ocean to boom and foam, and no brook to plash. Yet even these crust-cracking commotions, though they might be felt by the vibrations of the ground, would not manifest themselves audibly, for without air there can be no communication between the grating or cracking body and the nerves of hearing. Dead silence reigns on the moon: a thousand cannons might be fired and a thousand drums beaten upon that airless world, but no sound could come from them: lips might quiver and tongues essay to speak, but no action of theirs could break the utter silence of the lunar scene.

At a rate twenty-eight times slower than upon earth, the shadows shorten till the sun attains his meridian height, and then, from the tropical region upon which we have in imagination stood, nothing is to be seen on any side, save towards the black sky, but dazzling light. The relief of afternoon shadow comes but tardily, and the darkness drags its slow length along the valleys and creeps sluggishly up the mountain-sides till, in a hundred hours or more, the time of sunset approaches. This phenomenon is but daybreak reversed, and is unaccompanied by any of the gorgeous sky tints that make the kindred event so enrapturing on earth. The sun declines towards the dark horizon without losing one jot of its brilliancy, and darts the full intensity of its heat upon all it shines on to the last. Its disc touches the horizon, and in half an hour dips half-way beneath it, its intrinsic brightness and colour remaining unchanged. The brief interval of twilight occurs, as in the morning, when only a small chord of the disc is visible, and the long shadows now sharpen as the area of light that casts them decreases. For a while the zodiacal light vies with the earth-moon high in the heavens in illuminating the scene; but in a few hours this solar appendage passes out of view, and our world becomes the queen of the lunar night.

At this sunset time the earth, nearly in the zenith of us, will be at its half-illuminated phase, and even then it will shed more light than we receive upon the brightest of moonlight nights. As the night proceeds, the earth-phase will increase through the gibbous stages until at midnight it will be "full," and our orb will be seen in its entire beauty. It will perform at least one of its twenty-four-hourly rotations during the time that it appears quite full, and the whole of its surface features will in that time pass before the lunar spectator's eye. At times the northern pole will be turned towards our view, at times the southern; and its polar ice-caps will appear as bright white spots, marking its axis of rotation. If our lunar sojourn were prolonged we should observe the northern ice-cap creep downwards to lower latitudes (during our winter) and retreat again (during our summer); and this variation would be perceptible in a less degree at the southern pole, on account of the watery area surrounding it. The seas would appear (so far as can be inferred) of pale blue-green tint; the continents parti-coloured: and the tinted spots would vary with the changing terrestrial seasons, as these are indicated by the positions and magnitudes of the polar ice-caps. The permanent markings would be ever undergoing apparent modification by the variations of the white cloud-belts that encircle the terrestrial sphere. Of the nature of these variations meteorological science is not as yet in a position to speak: it would indeed be vastly to the benefit of that science if a view of the distribution of clouds and vapours over the earth's surface, as comprehensive as that we are imagining, could really be obtained.

It might happen at "full-earth," that a black spot with a fainter penumbral fringe would appear on one side of the illuminated disc and pass somewhat rapidly across it. This would occur when the moon passed exactly between the sun and the earth, and the shadow of the moon was cast upon the terrestrial disc. We need hardly say that these shadow-transits would occur upon those astronomically important occasions when an eclipse of the sun is beheld from the earth.

The other features of the sky during the long lunar night would not differ greatly from those to which we alluded in speaking of its day aspects. The stars would be the more brightly visible, from the greater power of the eye-pupil to open in the absence of the glaring sun, and on this account the milky-way would be very conspicuous and the brighter nebulæ would come into view. The constellations would mark the night by their positions, or the hours might be told off (in periods of twenty-four each) by the successive reappearances of surface features on certain parts of the terrestrial disc. The planets in opposition to the sun would now be seen, and a comet might appear to vary the monotony of the long lunar night. But a meteor would never flash across the sky, though dark meteoric particles and masses would continually bombard the lunar surface, sometimes singly, sometimes in showers. And these would fall with a compound force due to their initial velocity added to that of the moon's attraction. As there is no atmosphere to consume the meteors by frictional heat or break by its resistance the velocity of their descent, they must strike the moon with a force to which that of a cannon-ball striking a target is feeble indeed. A position on the moon would be an unenviable stand-point from this cause alone.

The lunar landscape by night needs little description: it would be lit by the earth-moon sufficiently to allow salient features, even at a distance, to be easily made out, for its moon (*i.e.* the earth) has thirteen times the light-reflecting area that ours has. But the night illumination will change in intensity, since the earth-moon varies from half-full to full, and again to half-full, between sunset and the next sunrise. The direction of the light, and hence the positions of the shadows, will scarcely alter on account of the apparent fixity of the earth in the lunar sky. A slight degree of warmth might possibly be felt with the reflected earthlight; but it would be insufficient to mollify the intensity of the prevailing cold. The heat accumulated by the ground during the three hundred hours' sunshine radiates rapidly into space, there being no atmospheric coat to retain it, and a cooling process ensues that goes on till, all warmth having rapidly departed, the previously parched soil assumes a temperature approaching that of celestial space itself, and which has been, as we have stated, estimated at about 200° below the Fahrenheit zero. If moisture existed upon the moon, its night-side would be bound in a grip of frost to which our Arctic regions would be comparatively tropical. But since there is no water, the aspect of the lunar scenery remains unmodified by effects of changing temperature.

Such, then, are the most prominent effects that would manifest themselves to the visual and other senses of a being transported to the moon. The picture is not on the whole a pleasant one, but it is instructive; and our rendering of it, imperfect though it be, may serve to suggest other inferences that cannot but add to the interest which always attaches to the contemplation of natural scenes and phenomena from points of view different from those which we ordinarily occupy.

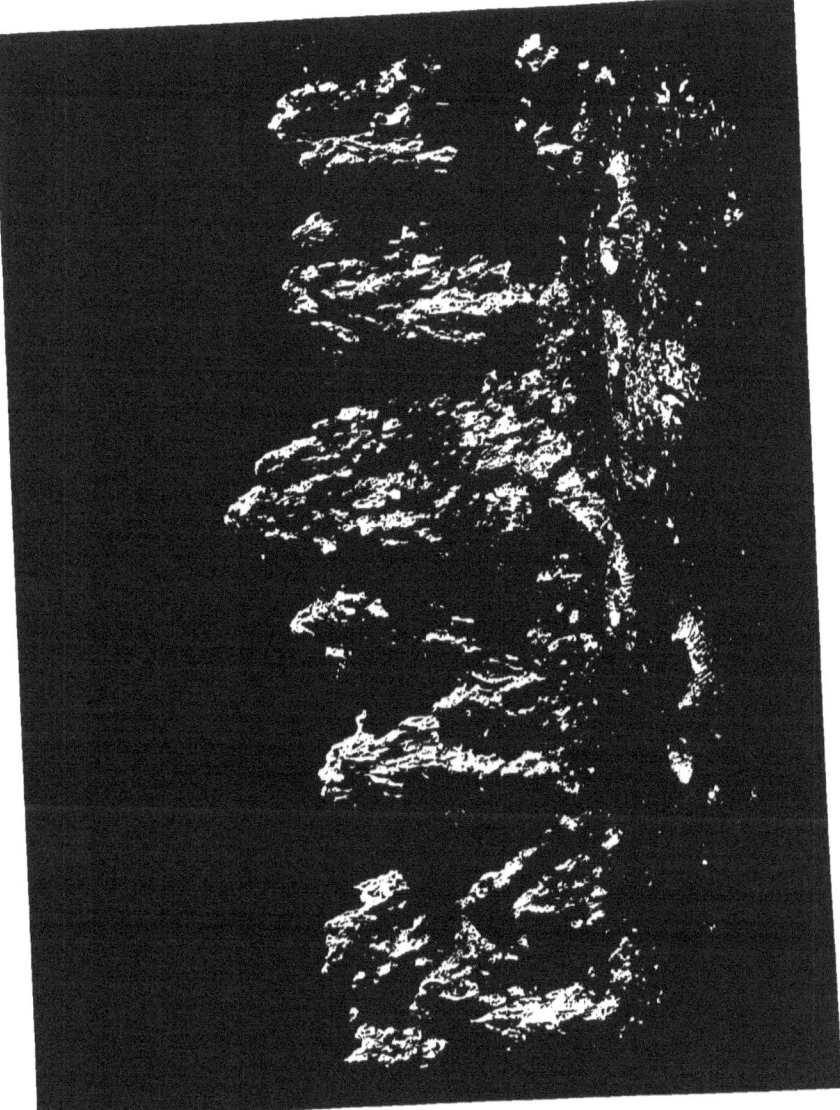

CHAPTER XIV.

THE MOON AS A SATELLITE: ITS RELATION TO THE EARTH AND MAN.

APART from the recondite functions of the moon considered as one of the interdependent members of the solar family, into which it would be beyond our purpose to inquire, there are certain means by which it subserves human interests and ministers to the wants of civilized man to which we deem it desirable to call attention, especially as some of them are not so self-apparent as to have attracted popular attention.

The most generally appreciated because the most evident of the uses of the moon is that of a luminary. Popular regard for it is usually confined to its service in that character, and in that character poets and painters have never tired in their efforts to glorify it. And obviously this service as a "lesser light" is sufficiently prominent to excite our warmest admiration. But moonlight is, from the very conditions of its production, of such a changeable and fugitive nature, and it affords after all so partial and imperfect an alleviation of night's darkness, that we are fain to regard the light-giving office of the moon as one of secondary importance. Far more valuable to mankind in general, so estimable as to lead us to place it foremost in our category of lunar offices, is the duty which the moon performs in the character of a sanitary agent. We can conceive no direful consequences that would follow from a withdrawal of the moon's mere light; but it is easy to imagine what highly dangerous results would ensue if the moon ceased to produce the tides of the ocean. Motion and activity in the elements of the terraqueous globe appear to be among the prime conditions in creation. Rest and stagnation are fraught with mischief. While the sun keeps the atmosphere in constant and healthy circulation through the agency of the winds, the moon performs an analogous service to the waters of the sea and the rivers that flow into them. It is as the chief producer of the tides—for we must not forget that the sun exercises *its* tidal influences, though in much lesser degree—that we ought to place the highest value on the services of the moon : but for its aid as a mighty scavenger, our shores, where rivers terminate, would become stagnant deltas of fatal corruption. Twice (to speak generally) a day, however, the organic matter which rivers deposit in a decomposing state at their embouchures is swept away by the tidal wave ; and thus, thanks to the moon, a source of direful pestilence is prevented from arising. Rivers themselves are providentially cleansed by the same means, where they are polluted by bordering towns and cities which, from the nature of things, are sure to arise on river banks ; and it seems to be also in the nature of things that the river traversing a city must become its main sewer. The foul additions may be carried down by the stream in its natural course towards the ocean, but where the river is large there will be a decrease in velocity of the current near the mouth or where it joins the sea, thus causing partial stagnation and consequent deposition of the deleterient matters. All this, however, is removed, and its inconceivable evils are averted by our mighty and ever active " sanitary commissioner," the moon. We can scarcely doubt that a healthy influence of less obvious degree is exerted in the wide ocean itself ; but, considering merely human interests, we cannot suppress the conviction that man is more widely and immediately benefited by this purifying office of the moon than by any other.

But the sanitary service is not the only one that the moon performs through the agency of the tides. There is the work of tidal transport to be considered. Upon tidal rivers and on certain coasts, notwithstanding wind and the use of steam, a very large proportion of the heavy merchandize is transported by that slow but powerful " tug " the flood-tide ; and a similar service, for which, however, the moon is not to be entirely credited, is done by the down-flow of the ebb-tide-

Large ships and heavily-laden rafts and barges are quietly taken in tow by this unobtrusive prime mover, and moved from the river's mouth to the far-up city, and from wharf to wharf along its banks; and a vast amount of mechanical work is thus gratuitously performed which, if it had to be provided by artificial means, would represent an amount of money value which for such a city as London would have to be counted by thousands, possibly millions, of pounds yearly. For this service we owe the moon the gratitude that we ought to feel for a direct pecuniary benefactor.

In the existing state of civilization and prosperity, we do not, however, utilize the power of the tides nearly to the extent of their capabilities. Our coal mines, rich with the "light of other days" —for coal was long ago declared by Stevenson to be "bottled sunshine"—at present furnish us with so abundant a supply of power-generating material that in our eagerness to use it upon all possible occasions we are losing sight, or putting out of mind, many other valuable prime movers, and amongst them that of the rise and fall of the waters, which can be immediately converted into any form of mechanical power by the aid of tide-mills. Such mills may be found in existence here and there, but for the present they are generally out-rivalled by the steam engine with all its conveniences and adaptabilities; and hence they have not shared the benefits of that inventive ingenuity which has achieved such wonders of mechanical appliance while steam has been in the ascendant. But it must be remembered that in our extravagant use of coal we are drawing from a bank into which nothing is being paid. We are consuming an exhaustive store, and the time must come when it will be needful to look around in quest of "powers that may be." Then an impetus may be given to the application of the tides to mechanical purposes as a prime mover.* For the people of the British Islands the problem would have an especial importance, viewing the extent of our seaboard and the number of our tidal rivers. The source of motion that offers itself is of almost incalculable extent. There is not merely the onward flowing

* About 100 years ago London was supplied with water chiefly by pumps worked by tidal mills at London Bridge.

motion of streams to be utilized, but also the *lift* of water, which, if small in extent, is stupendous in amount; and within certain limits it matters little to the mechanician whether the "foot-pounds" of work placed at his disposal are in the form of a great mass lifted to a small height or a small mass lifted to a great height. There is no reason either why the utilization of the tides should be confined to rivers. The sea-side might well become the circle of manufacturing industry, and the millions of tons of water lifted several feet twice daily on our shores might be converted, even by schemes already proposed, to furnish the prime movement of thousands of factories. And we must not forget how completely modern science has demonstrated the inter-convertibility of all kinds of force, and thus opened the way for the introduction of systems of transporting power that, in such a state of things as we are for the moment considering, might be of immense benefit. Gravity, for instance, can be converted into electricity; and electricity gives us that wonderful power of transmitting *force* without transmitting (or even moving) *matter*, which power we use in the telegraph, where we generate a force at one end of a wire and *use* it to ring bells or deflect needles at the other end, which may be thousands of miles away. What we do with the slight amount of force needful for telegraphy is capable of being done with any greater amount. A tide-mill might convert its mechanical energy by an electro-magnetic engine, and in the form of electricity its force could be conveyed inland by proper wires and there reconverted back to mechanical or moving power. True, there would be a considerable loss of power, but that power would cost nothing for its first production. Another means ready to hand for transporting power is by compressed air, which has already done good service; another is the system so admirably worked out by Sir W. Armstrong, of transmitting water-power through the agency of an "accumulator," now so generally used at our Docks and elsewhere for working cranes and such other uses. And as the whole duty of the engineer is to *convert* the forces of nature, there is a rich field open for his invention, and upon which he may one day have to enter, in adapting the pulling force of the moon to his fellow man's mechanical wants through the intermediation of the tides.

THE MOON AS A SATELLITE.

Another of the high functions of the moon is that by which she subserves the wants of the navigator, and enables him to track his course over the pathless ocean. Of the two co-ordinates, Latitude and Longitude, that are needful to determine the position of a ship at sea (or of any standpoint upon the earth's surface) the first is easily found, inasmuch as it is always equal to the altitude of the celestial pole at the place of observation. But the determination of the longitude has always been a difficult problem, and one upon which a vast amount of ingenuity has been expended. When it was first attacked it was soon discovered that the moon was the object of all others by which it could be most accurately and, all things considered, most readily determined. We must premise that the longitude of one place from another is in effect the difference between the local times at the two places, so that when we say that a place or a ship is, for instance, seven hours, twenty-four minutes, ten seconds, west of Greenwich, we mean that the time-o'-day at the place or ship is seven hours twenty-four minutes ten seconds earlier than that at Greenwich. Hence, finding the longitude at sea or at any place and moment means finding what time it is at Greenwich at that moment. Of course this could be most easily done if we could set a timekeeper at Greenwich and rely upon its keeping time during a long sea voyage; and this plan appeared so feasible that our Government long ago offered a prize of £20,000 for a timekeeper which would perform to a stated degree of accuracy after a certain sea voyage. One John Harrison did make such a timekeeper, that actually satisfied the conditions, and obtained the prize: and chronometers are now largely used for longitude, their construction having been brought to great perfection, especially in England, owing to a continuance (in a less liberal degree, however) of Government inducement. But chronometers are not entirely to be relied on, even where several are carried, which in other than Government ships is rarely the case: recourse must be had to the heavenly bodies for check upon the timekeeper. And the moon is, as we have said, the body that best serves the requirements of the problem.

The lunar method for longitude amounts practically to this. The stars are fixed; the sun, moon, and planets move amongst them; the sun and planets with very slow rates of apparent motion, the moon with a very rapid one. If, then, it be predicted that at a certain instant of Greenwich time the moon will be a certain distance from a fixed star, and if the mariner at sea observes *when* the moon has that exact distance, he will know the Greenwich time at the instant of his observation.* The moon thus becomes to him as the hand of a timepiece, whereof the stars are the hour and minute marks, the whole being, as it were, set to Greenwich time. The requisite predictions of the distance (as seen from the earth's centre) of the moon from convenient fixed stars, or from the sun, or any of the principal planets—whose calculated places are so accurate that they may for this purpose be used as fixed stars—are given to the utmost exactness in the navigators' *vade mecum*, the "Nautical Almanac," for every third hour, day and night, of Greenwich time (except for a few days near new-moon, when the moon cannot be seen); and from these given distances the navigator can, by a simple process of differencing, obtain the Greenwich time corresponding to the distance which he may have observed.† Then knowing, as he does by other observations easily obtained, the local or ship's time of his observation, he takes the difference between this and the corresponding Greenwich time, and this difference is his longitude from Greenwich. Of course the whole value of this method depends upon the exactitude of the predicted distances corresponding to the given Greenwich times. These distances are obtained by tables of the moon's motions, which must be found from observations. The motions in question are of an intricacy almost past comprehension, on account of the disturbing forces to which the moon is subjected by the sun and planets. The powers of the profoundest mathematicians, from Newton downwards, have been severely exercised in efforts to group them into a theory, and represent them by tables capable of furnishing the requisite exact predictions of lunar positions for nautical purposes. Accurate observations of the moon's place night after night have,

---

* The sun and planets are comparatively useless for this object, because of their slow movement among the stars; the change of their positions from hour to hour is so small as to render uncertain the Greenwich times deducible therefrom. Their use would be comparable to taking the time from the hour-hand of a clock.

† Certain corrections are necessary to clear his observed distance of the effects of parallax and refraction; upon these, however, we cannot enter here.

from the dawn of this lunar method for longitude, been in urgent request by mathematicians for the purposes specified, and it was solely to procure these observations that the Observatory at Greenwich was established, and mainly for their continued prosecution (and for the stellar observations necessary for their utilization) that it is sustained. For two centuries the moon has been unremittingly observed at Greenwich, and the tables at present used for making the "Nautical Almanac" (those formed by Prof. Hansen) depend upon the observations there obtained. The work still goes on, for even now the degree of exactitude is not what is desired, and astronomers are looking forward with some interest to new lunar tables which were left complete by the late M. Delaunay, formerly the head of astronomy in France, based upon a theory which he evolved. This use of the moon is the grandest of all in respect of the results to which it has led.

Then, too, regarding the moon as a timekeeper, we must not forget the service that it renders in furnishing a division of time intermediate between the day—which is measured by the earth's rotation—and the year, which is defined by the earth's orbital revolution. Notwithstanding the survival of lunar reckoning in our religious services, we, in our time and country, scarcely need a moon to mark our months; but we must not forget that with many ancient people the moon was, and with some is still, the chief timekeeper, the calendars of such people being lunar ones, and all their events being reckoned and dated by "moons." To us, however, the moon is of great service in this department by enabling us to fix dates to many historical events, the times of occurrence of which are uncertain, by reason of defective records or by dependence upon such uncertain data as "lives of emperors," years of this or that king's reign, or generations of one or another family. The moon now and then clears up a mystery, or decides a disputed point in chronology, by furnishing the accurate date of an ancient eclipse, which was a phenomenon that always inspired awe and secured for itself careful record. The chronologer is continually applying to the astronomer for the date and place of visibility of some total eclipse, of which he has found an imperfect record, veritable as to the fact, but dated only by reference to some year of a so-and-so's reign, or by some battle or other historical occurrence. The eclipses that occurred near the time are then examined, and when one is found that tallies with recorded conditions in other respects (such as the time of day and the place of observation), its indisputable date becomes a starting-point from which the chronologer works backwards and forwards in safety. There is one famous eclipse—that predicted by Thales six centuries before Christ, which put an end to the battle between the Medes and Lydians by the terror its darkness created in both armies—which is most intimately associated with ancient chronology, and has been used to rectify a proximate date (the first year of Cyrus of Babylon) which forms the foundation of all Scripture chronology. Sacred and profane history alike are continually receiving assistance from the accurate dates which the moon, by having caused eclipses of the sun, enables the astronomer to fix beyond cavil or doubt.

The mention of eclipses reminds us, too, of the use which the moon has been in increasing, through them, our knowledge of the physical condition of the sun. If the moon had never intervened to cut off the blinding glare of the solar disc, we should have been to this day left to assume that the sun is all-contained by the dazzling globe that we ordinarily see. But, thanks to the moon's intervention, we now know that the sun is by no means the mere naked sphere we should have suspected. Eclipses have taught us that it is surrounded by an envelope of glowing gases, and that it has a vast vaporous surrounding, beyond its glowing atmosphere, which appears to be composed of matter streaming away from the sun into surrounding space. With these discoveries still in their infancy, it is impossible to foresee the knowledge to which they will eventually lead, but they can hardly be barren of fruit, and whatever they ultimately teach will be so much insight gained into the sublimest problem that human science has before it—the determination of the source and maintaining power of the light and heat and vivifying agency of the sun. In according our thankful reflections to the moon for these revelations, we must not forget that, should there be inhabitants upon our neighbouring worlds, Mercury, Venus, and Mars, which have no satellites, they, the supposed inhabitants, can gain no such knowledge upon the surroundings of the ruler of the solar system. On the other hand, any rational being who may be supposed to dwell upon Saturn or Jupiter, would, through the intervention of their numerous moons, have, in the latter case especially, far more abundant opportunities of acquiring the knowledge in question than we have.

Finally, there is a use of the moon which touches us, author and reader, very closely. It has taught us of a world in a condition totally different from our own; of a planet without water, without air, without the essentials to life development, but rather with the conditions for life destruction; a planet left by the Creator—for wise purposes that we cannot fully know—as it were but half-formed, with all the igneous foundations fresh from the cosmical fire, and with its rough-cast surface in its original state, its fire and mould-marks exposed to our view. From these we have essayed to resolve some of the processes of formation, and thus to learn something of the cosmical agencies that are called forth in the purely igneous era of a planet's history. We trust that we, on our part, have shown that the study of the moon may be a benefit not merely to the astronomer, but to the geologist; for we behold in it a mighty "medal of creation" doubtless formed of the same material and struck with the same die that moulded our earth; but while the dust of countless ages and the action of powerful disintegrating and denuding elements have eroded and obliterated the earthly impression, the superscriptions on the lunar surface have remained with their pristine clearness unsullied, every vestige sharp and bright as when it left the Almighty Maker's hands. The moon serves no second-rate or insignificant service when it teaches us of the variety of creative design in the worlds of our system, and exalts our estimation of this peopled globe of ours by showing us that all the planetary worlds have *not* been deemed worthy to become the habitations of intelligent beings.

Reflections upon the uses of the moon not unnaturally lead our thoughts to some matters that may be regarded as abuses. These mainly take the form of superstitions, erroneous beliefs in the moon's influence over terrestrial conditions, and occasionally of erroneous ideas upon the moon's functions as a luminary. The first-mentioned are almost beneath notice, for they include such mythical suspicions as that the moon influences human sanity and other affections of mind and body; that the moon's rays have a decomposing effect upon organic matter; that they produce blindness by shining upon a sleeper's eyes; that the moon determines the hours of human death, which is supposed to occur with the change of the tide, etc. All such, having no foundation on fact, are put beyond our consideration. The third matter we have mentioned may also be dismissed in a very few words. The erroneous ideas upon the moon's functions as a luminary, to which we allude, are those which are manifested by poets and painters, and even historians, who do not hesitate to bring the moon upon a scene in any form and at any time they please without reference to actual lunar circumstances. It is no uncommon thing to see, in a picture representing an evening scene, a moon introduced which can only be seen in the morning—a waning moon instead of a waxing one; and astronomical critics have, indeed, caught artists so far tripping as to put a moon in a picture representing some event that occurred upon a date when the moon was new, and therefore invisible. Writers take the same liberties very frequently. A newspaper correspondent, during the Franco-Prussian war, described the full moon as shining upon a scene of desolation on a particular night, when really there was no moon to be seen. One of the most flagrant cases of this kind, however, occurs in Wolfe's ballad on "The death of Sir John Moore," where it is written that the hero was buried "By the struggling moonbeam's misty light." But the interment actually took place at a time when the moon was out of sight. We mention these abuses of the moon in the hope of promoting a better observance of the moon's luminary office. They who wish to bring the moon upon a scene, not knowing *ipso facto* that it was there, should first take the advice of Nick Bottom in the "Midsummer Night's Dream," and make sure of their object by consulting an almanac.

The second of the specified abuses to which the moon is subject refers to its supposed influence on the weather; and in the extent to which it goes this is one of the most deeply rooted of popular errors. That there is an infinitesimal influence exerted by the moon on our atmosphere will be seen from the evidence we have to offer, but it is of a character and extent vastly different from what is commonly believed. The popular error is shown in its most absurd form when the mere *aspect* of the moon, the mere transition from one phase of illumination to another, is asserted to be productive of a change of weather; as if the gradual passage from first quarter to second quarter, or from that to third, could of itself upset an existing condition of the atmosphere; or as if the conjunction of the moon with the sun could invert the order of the winds, generate clouds, and pour down rains. A moment's reasoning ought to show that the supposed cause and the observed effect have no necessary connection. In our climate the weather may be said to change at least every three days, and the moon changes—to retain the popular term—every seven days; so that the probability of a coincidence of these changes is very great indeed: when it occurs, the moon is sure to be credited with causing it. But a theory of this kind is of no use unless it can be shown to apply in every case; and, moreover, the change must always be in the same direction; to suppose that the moon can turn a fine day to a wet one, and a wet day to a fine morrow indiscriminately, is to make our satellite blow hot and cold with the same mouth, and so to reduce the supposition to an absurdity. If any marked connection existed between the state of the air and the aspect of the moon, it must inevitably have forced itself unsought upon the attention of meteorologists. In the weekly return of Births, Deaths, and Marriages, issued by the Registrar-General, a table is given, showing all the meteorological elements at Greenwich for every day of the year, and a column is set apart for noting the changes and positions of the moon. These reports extend backwards nearly a quarter of a century. Here, then, is a repertory of data that ought to reveal at a glance any such connection, and would certainly have done so had it existed. But no constant relation between the moon columns and those containing the instrument readings has ever been traced. Our meteorological observatories furnish continuous and unbroken records of atmospheric variations, extending over long series of years: these afford still more abundant means for testing the validity of the lunar hypothesis. The collation has frequently been made for special points in the inquiry, and certainly *some* connection has been found to obtain between certain positions of the moon in her orbit and certain instrumental averages; but so small are the effects traceable to lunar influence, that they are almost inappreciable among the grosser irregularities that arise from other and as yet unexplained causes.

The lunar influences upon our atmosphere most likely to be detected are those of a tidal character, and those due to the radiation of the heat which the moon receives from the sun. The first would be shown by the barometer, which may be called an "atmospheric tide gauge." Some years ago Sir Edward Sabine instituted a series of observations at St. Helena, to determine the variations of barometric indications from hour to hour of the lunar day. The greatest differences were found to occur between the times when the moon was on the meridian, and when it was six hours away from the meridian; in other words, between atmospheric high tide and low tide. But the average of these differences amounted only to the four-hundredth part of an inch on the instrument's scale; a quantity that no weather observer would heed, that none but the best barometers would show, and that can have no perceptible effect on weather changes. The distance of the moon from the earth varies, as is well known, in consequence of the elliptical form of her orbit: this variation ought also to produce an effect upon the instrument's indications; but Colonel Sabine's analysis showed that it was next to insensible; the mean reading at apogee differing from that at perigee by only the two-thousandth part of an inch. Schubler, a German meteorologist, had arrived at similarly negative results some years previously. Hence it appears that the great index of the weather is not sensibly affected by the state of the moon; the conclusion to be drawn with regard to the weather itself is obvious enough. As regards the heat received from the moon, we know, from the recent experiments of Lord Rosse in England, and Marie Davy in France, elsewhere alluded to, that a degree of warmth appreciable to the highly sensitive thermopile is exerted by the moon upon the earth near to the time of full moon, when the sun's rays have been pouring their unmitigated heat upon the lunar surface continuously for fourteen days. And as it is improbable that the whole of the heat sent earthwards from the moon reaches the earth's surface, we must infer that a considerable amount is absorbed in the higher atmosphere, and does work in evaporating the lighter clouds and thinning the denser ones. The effect of this upon the earth is to facilitate the radiation of its heat into space, and so to cool the lower atmospheric strata. And this effect has been shown to be a veritable one by an exhaustive tabulation of temperature records from various observatories, which was undertaken by Mr. Park Harrison. The general conclusion from these was, that the temperature at the earth's surface is lower by about $2\frac{1}{2}$ degrees at moon's last quarter than at first quarter; the paradoxical result being what would naturally follow from the foregoing consideration. The tendency of the full moon to clear the sky has been remarked by several distinguished authorities, to wit, Sir John Herschel, Humboldt, and Arago; and in general the clearing may be accepted as a meteorological fact, though in one case of close examination it has been negatived. It cannot be doubted that a full moon sometimes shows a night to be clear that would in the absence of the moon be called cloudy.

When close comparisons are made between the moon's positions and records of rain-fall and wind-direction, dim indications of relation exhibit themselves, which may be the feeble consequences of the change of temperature just spoken of; but in every case where an effect has been traced it has been of the most insignificant kind, and no apparent connexion has been recognized between one effect and another. Certainly there is nothing that can support the extensive popular belief in lunar influence on weather, and nothing that can modify the conviction that this belief as at present maintained is an absurd delusion. Yet its acceptance is so general, and runs through such varied grades of society, that we have felt it our duty to dwell upon it to the extent that we have done.

CHAPTER XV.

CONCLUDING SUMMARY.

HAVING arrived at the conclusion of our subject, it appears to us desirable that we should recall to the reader, by a rapid review, its salient features.

Our main object being to attempt what we conceive to be a rational explanation of the surface details of the moon which should be in accordance with the generally received theory of planetary formation, and with the peculiar physical conditions of the lunar globe—the opening of our work was a summary of the nebular hypothesis as it was started by the first Herschel and systemised by Laplace. Following these philosophers we endeavoured to show how a chaotic mass of primordial matter existing in space would, under the action of gravitation, become transformed into a system of planetary bodies circulating about a common centre of gravity; and further, how, in some cases, the circulating planetary masses would themselves become sub-centres of satellitic systems; our earth being one of these sub-centres with only one satellitic attendant—to wit, the moon, the subject of our study.

The moon being thus considered as evolved from the parent nebulous mass, and existing as an isolated and compact body, we had next to consider what was the effect of the continued action of the gravitating force. By the light of the beautiful "mechanical theory of heat" we argued that this force, not being *destructible*, but being *convertible*, was turned into heat; and that whatever may have been the original condition of the parent nebulous mass, as regards temperature, its planetary offspring became elevated to an intense degree of heat as they assumed the form of spheres under the influence of gravitation.

The incandescent sphere having attained its maximum degree of heat by the total conversion thereinto of the gravitating force it embodied, we explained how there must have ensued a dispersion of that heat by radiation into surrounding space, resulting in the cooling and consequent solidification of the outermost stratum of the lunar sphere, and subsequently in the continuation of the cooling process downwards or inwards to the centre. And here we essayed to prove that in this second stage of the cooling process, when the crust was solid and the subjacent portion of the molten sphere was about to solidify, there would come into operation a principle which appears to govern the behaviour of certain fusible substances, and which may be concisely termed the principle of pre-solidifying expansion. We adduced several examples of the manifestation of this principle, soliciting for it the careful consideration of physicists and geologists, and looking to it as furnishing the key to the mystery of volcanic action upon the moon, since, without needing recourse to aqueous or gaseous sources of eruptive power, it afforded a rationale of the ejection of the fluid and semifluid matter of the moon through the solidified crust thereof, and also of the dislocations of that crust, unattended by actual ejection of subsurface matter, of which our satellite presents a variety of examples, and which the earth also appears to have experienced at some period of its formative history.

Arrived at this stage of our subject we thought it needful to introduce some pages of data and descriptive detail. Accordingly in one chapter we discussed the form, magnitude, weight, and density of the moon, and the force of gravity at its surface : and the more soundly to fix these data in the mind, we devoted a few lines to explanation of the methods whereby each has been ascertained. We then examined the question (so important to our subject) of the existence or non-existence of a lunar atmosphere, giving the evidence, which may be regarded as conclusive, in proof of the absence of both air and water from the moon, and, therefore, refuting the claim of these elements to be considered as sources or influants of the moon's volcanic manifestations. A general *coup d'œil* of the lunar hemisphere facing the earth next engaged our attention, and we considered the aspect of the disc as it is viewed by the naked eye and with telescopes of various powers. From this general survey we passed to the topography of the moon, tracing briefly the admirable labours of those who have advanced this subject, and, by aid of picture and skeleton maps and a table of position co-ordinates, placing it within the reader's power to become more than sufficiently acquainted for the purposes of this work with the names and positions of detailed objects and features of interest. Special descriptions of interesting and typical spots and regions were given in some few cases where such appeared to be called for.

These descriptive matters disposed of, we proceeded to discuss the various classes of surface features with a view to explaining the precise actions which appear to us to have led to their formation. Naturally the craters first demanded our attention. We pointed out the reasons for regarding the great majority of the circular formations of the moon as craters, as truly volcanic as those of which we have examples, modified by obvious causes, upon the earth; and, tracing the causative phenomena of terrestrial volcanoes, we showed how the explanations which have been offered to account for them scarcely apply to those of the moon: and thus, driven to other hypotheses, we endeavoured to demonstrate the probability of the lunar craters having been produced by eruptive force, generated by that pre-solidifying expansion of successive portions of the moon's molten interior, which we enunciated in our third chapter. The precise course of phenomena which resulted in the production of a crater of the normal lunar type, with or without the significant central cone, were then illustrated by a series of step-by-step diagrams with accompanying descriptive paragraphs. And after treating of craters of the normal type we pointed out and explained some variations thereupon that are here and there to be met with, and likewise those curious complications of arrangement which exhibit craters superimposed one upon another and intermingled in strange confusion.

From craters manifestly volcanic we passed to the consideration of those circular formations which, from their vastness of size, scarcely admit of satisfactory explanation by a volcanic hypothesis. We summarized several proffered theories of their origin, and pointed out what we considered might be a possible key to the solution of the selenological enigma which they constitute, without, however, expressing ourselves entirely satisfied with the validity of our suggestion. The less mysterious features presented by peaks and mountain ranges were then discussed to the extent that we considered requisite, viewing their comparatively simple character and the secondary position they occupy in point of numerical importance upon the moon. At greater length we dealt with the cracks and chasms and the allied phenomena of radiating streaks, pointing out with regard to these latter the strikingly beautiful correspondence in effect (and therefore presumably in cause) between them and crack-systems of a glass globe "starred" by an expanding internal medium.

The more notable objects and features of the lunar surface being disposed of, we had next to say a few words upon some residual phenomena, chiefly upon the colour of lunar surface details, and upon their various degrees of brightness or reflective power. And, inasmuch as varying brightness seemed to us to be related to varying antiquity, we were thence led to the question of the chronology of selenological formations, and to the disputation upon the continuance of volcanic action upon the moon in recent years. We regarded this question from the observational and the inferential points of view, and were led to the conclusion that the moon's surface arrived at its terminal condition ages ago, and that it is next to hopeless to look for evidence of existing change.

Thus far our work dealt with the moon as a planetary body merely. It occurred to us, however, that we might add to the interest attaching to our satellite were we to regard it for a time as a world, and consider its conditions as respects fitness for habitation by beings like ourselves. The arguments against the possibility of the moon being thus fitted for human creatures, or, indeed, for any high organism, were decisive enough to require little enforcing. It appeared to us, nevertheless, that much might be learnt by imagining one's self located upon the moon during a period embracing one lunar day (a month of our reckoning), with power to comprehend the peculiar circumstances and conditions of such a situation. We therefore attempted a description of an imaginary sojourn upon the moon, and pointed out some of the more striking aspects and phenomena which we know by legitimate inference would be there manifested. We trust, that while our modest efforts in the chapter referring to this branch of our subject may prove in some degree entertaining, they may be in a greater degree instructive, inasmuch as certain facts are brought into prominence which would not unnaturally be overlooked in contemplating the moon from the earth, the only *real* stand-point that is available to us.

In our final chapter we considered the moon as a satellite, and sought to enhance popular regard for it on account of certain high functions which it performs for man's benefit on this earth ; but which are in great risk of being overlooked. We showed that, notwithstanding the moon's occasionally useful service as a nocturnal luminary, it fills a far higher office as a sanitary agent by cleansing the shores of our seas and rivers through the agency of the tides. We pointed out the vast amount of absolutely mechanical work and commercial labour which the same tidal agency executes in transporting merchandize up and down our rivers—an amount that, to take the port of London alone, represents a money value *per annum* that may be reckoned in millions sterling, seeing that if our river was tideless all transport would have to be done by manual or steam power. We then hinted at the stupendous reservoir of power that the tidal waters constitute, a form of power which has not as yet been sufficiently called into operation, but which may be invoked by-and-by, when we have begun to feel more acutely the consequences of our present prodigal use of the fuel that was stored up for us by bountiful nature ages upon ages ago. The moon's services to the navigator, in affording him a ready means of finding his longitude at sea ; to the chronologist and historian, as a timekeeper, counting periods too vast for accurate reckoning by other means ; to the astronomer and student of nature, in revealing certain wonderful surroundings of the solar globe, which, but for the phenomena of eclipses caused by the moon's interposition, would never have been suspected to exist—these were other functions that we dwelt upon, all too briefly for their deserts ; and, lastly, we spoke of the moon as a medal of creation fraught with instructive suggestions, which it has been our endeavour to bring to notice in the course of this work. And from uses we passed to abuses, directing attention to a few popular errors and widespread illusions relating to lunar influence upon and in connection with things terrestrial. This part of our work might have been considerably expanded, for, in truth, the moon has been a misunderstood and misjudged body. Some justice we trust we have done to her : we have brought her face to the fireside; we have analysed her features, and told of virtues that few of her admiring beholders conceived her to possess. We have traced out her history, fraught with wonderful interest, and doubtless typical of the history of other spheres that in countless numbers pervade the universe: and now, having done our best to make all these points familiar, we commend the moon to still further study and still more intimate acquaintance, confident that she will repay all attentions, be they addressed to her as

A PLANET, A WORLD, OR A SATELLITE.

THE END.

www.ingramcontent.com/pod-product-compliance
Lightning Source LLC
Chambersburg PA
CBHW031351230426
43670CB00006B/507